Good Style

Good Style

Writing for science and technology

John Kirkman

Consultant on scientific and technical communication

E & FN SPON

An Imprint of Chapman & Hall

London · New York · Tokyo · Melbourne · Madras

Published by E & FN Spon, an imprint of Chapman & Hall, 2–6 Boundary Row, London SE1 8HN

Chapman & Hall, 2–6 Boundary Row, London SE1 8HN, UK

Chapman & Hall, 29 West 35th Street, New York NY10001, USA

Chapman & Hall Japan, Thomson Publishing Japan, Hirakawacho Nemoto Building, 7F, 1-7-11 Hirakawa-cho, Chiyoda-ku, Tokyo 102, Japan

Chapman & Hall Australia, Thomas Nelson Australia, 102 Dodds Street, South Melbourne, Victoria 3205, Australia

Chapman & Hall India, R. Seshadri, 32 Second Main Road, CIT East, Madras 600 035, India

First edition 1992

© 1992 John Kirkman

Typeset in 10/12pt Palatino by Graphicraft Typesetters, Hong Kong
Printed in Singapore by Fong & Sons Printers Pte. Ltd.

ISBN 0 419 17190 8

A catalogue record for this book is available from the British Library

Library of Congress Cataloging-in-Publication data available

Contents

Contents

Preface

In seminars and workshops over the past 30 years, I have had the good fortune to be able to discuss style with several thousand practising engineers and scientists. This book presents the essence of our discussions and some supporting statistical evidence.

I wish to acknowledge my debt to the participants in those seminars, who listened to my ideas, tried them out, and told me which were useful and which were not. I hope this presentation of the 'useful' category will contribute to the gradual improvement of standards of communication in science and technology.

I am well aware that pure scientists, applied scientists, engineers, and computer specialists are distinctive groups; but it would have been tedious for readers if I had referred continually to all four groups, or even just to 'engineers and scientists'. So, since the communication problems confronting professionals in engineering, science, and computing are very similar, and since this book is concerned with how to write readably, I hope readers will accept my adoption of the convenient all-embracing terms 'technical writers' and 'technical writing'. I have distinguished between groups of writers and types of writing occasionally, when there seemed an important reason for doing so.

This book is a revised edition of *Good Style for Scientific and Engineering Writing*, published by Pitman in 1980. The text has been substantially revised, and extended by about 50%. A principal aim in the extension has been to give greater attention to style for writing computer-related texts.

The principal revisions and additions are:

- a new section that describes a survey of responses to varying styles for computer manuals;
- a new chapter on avoiding 'distorted' English in computer-related texts;
- a new section on second-person style (using *you* and *your*);
- a new section on tone, with special emphasis on tone in on-screen text;
- a new section on 'fashionable' words;
- a new section on 'nominalization';

- a new section on use of nouns as pre-modifiers;
- an extension of the discussion of punctuation.

John Kirkman
Ramsbury
Wiltshire

REFERENCES

References are numbered in sequence throughout the book. Bibliographic details of each reference are given at the end of the book, starting on page 217.

CHAPTER ONE

Which style do technical readers prefer?

It is surely axiomatic that the aim of technical writing is to transmit information accurately, quickly and economically from one person to another. Then why do so many scientists and engineers make their writing so heavily unreadable?

Obviously, their subject matter is sometimes complex and conceptually difficult; but frequently the 'unreadability' stems from use of a style that makes the reader's task much heavier than it need be.

Scientists and engineers themselves complain about the heaviness of this style; but when I suggest that passive, impersonal, turgid expression is a millstone that the technical content need not carry, I am told that papers written in any other style would be unacceptable: 'It would be thrown straight back'; 'My boss wouldn't have it'; 'Editors insist that you write passively and impersonally'; 'You must make your work sound impressive'.

Always there is anxiety that other engineers and scientists would not accept a departure from 'traditional' style. This seems to be true whether I am talking to young men and women, or to their seniors in higher management positions. The anxiety is always caused by a spectral reactionary group 'they'.

I have believed for a long time that 'they' must really be a small minority. In hundreds of round-table discussions with individuals and groups of scientists and engineers, I have found wide agreement on what constitutes the most readable, the most effective style for scientific and technical writing. But this agreement has been hedged round with anxiety about what other readers will want or expect. So this book offers not only advice but also evidence to convince anxious writers that other engineers and scientists, 'bosses', 'editors' – 'they' – will accept, even prefer, the style advocated here.

To obtain evidence of preferences for styles, I have run four surveys, asking readers to give their views on six different versions of a scientific or technical text. In each survey, the subject-matter of the six versions was held constant; only the style of writing was different. Readers were asked to read the versions in any order, and then to say which they found most comfortable to read, easiest to grasp, and simplest to digest. They were also invited to comment on why they liked some versions and rejected others.

To obtain evidence of the style preferred in reports and papers, I ran three of the surveys with members of the Institution of Chemical Engineers, the British Ecological Society, and the Biochemical Society. In all three surveys, the majority of respondents voted for direct, active, judiciously personalized writing. Active style had majorities of 28%, 41% and 18.5% over 'traditional' passive style.

To establish what style is preferred in technical manuals, I gave readers six versions of part of a manual describing a computer program. Again, the majority of respondents voted for direct, active, judiciously personalized writing. Active, personalized style had a majority of 27.8% over a traditional impersonal style.

The versions on which the surveys were based, the explanatory notes that accompanied them, and the responses from readers are presented in the Appendices.

If you are just beginning to think about style for scientific and technical writing, I recommend that you read through the appendices before your read the main text of the book. In that way, you will be able to begin your consideration of style by getting a sense of the effects produced by varying choices of vocabulary and structure. In particular, I recommend teachers of writing to invite students to respond to the versions of texts in my surveys **before** analysing in detail the effects created by manipulating individual 'linguistic variables'. (I should be interested to hear results from any groups that are 'tested'.)

However, if you are well aware of the difficulties caused by various types of unskilful writing, you may prefer to carry straight on to Chapter 2. There you will find a detailed discussion of the tactical decisions that must be made to produce readable, effective scientific and technical texts.

CHAPTER TWO

On style in general

2.1 STYLE AS CHOICE

Style in writing is concerned with choice. Every writer has available the enormous resources of a whole language. English presents a particularly large range of choices of individual words, and of combinations of words into small and large 'structures' – idioms, phrases, clauses, sentences, paragraphs, sections, chapters. The choices we make create the 'style', which is a term covering balance, emphasis and tone.

There is no such thing as the 'correct' way of expressing any idea, fact or opinion. Each writer selects the arrangement of words that he or she thinks will best express the intended meaning; each writer selects the arrangement that he or she thinks will best give the balance, emphasis and tone necessary to produce the desired response from readers.

For example, it is possible to contend that the following sentences convey roughly the same information – the wish of the writer to warn readers not to open Valve X before it has cooled, because of the possibility of a dangerous explosion:

Valve X is not to be opened before cooling to 18°C because of the possibility of flash-back.

Do not open Valve X before it has cooled to 18°C; there is a danger of flash-back.

Owing to the danger of flash-back, Valve X must not be opened before it has cooled to 18°C.

On no account should you open Valve X before it cools to 18°C or you may well cause flash-back.

If you open Valve X before it cools to 18°C, there will be an almighty flash-back.

Danger of flash-back: do not open Valve X before it cools to 18°C.

The linguistic choices made here alter the emphasis and tone of the statements, though the meaning stays the same. The factors that make one or other of the versions more effective as written communication are not matters of accuracy or clarity of technical content, nor are they matters of grammar: they are bound up with the relationship between the writer and the reader(s), and with the context in which the exchange takes place.

In other words, in judging effectiveness of communication and suitability of style, we must take into account both the **accuracy** and the **propriety** of the language chosen. If this book is to offer advice on the best style for scientific and technical writing, I must begin by defining the type(s) of writing I have in mind, by defining the writing/reading contexts in which the exchanges are to take place. Only then will it be possible to discuss which tactical choices will best convey the desired meaning with the accuracy, balance, emphasis, and tone required.

Broadly, I am concerned with the types of writing that are used for passing scientific and technical information between professional staff in academic, industrial and research organizations; that is, between groups whose intellectual capacities are approximately equal but whose specialist backgrounds may be very different. I am concerned, therefore, with reports, journal articles, proposals, technical memoranda, operating instructions and procedures, specifications, user's guides, reference manuals, and support documentation for equipment.

I am not suggesting tactics for explaining complex scientific and technical ideas to the general public. Mass communication poses special problems of selection and presentation that are not the concern of the average engineer, scientist or technical writer. Nor am I suggesting tactics for advertising and commercial writing. Though **all** communication is in a sense an attempt to sell someone else our ideas, there is a difference in degree between the tactics needed to advocate a change from one soap powder to another, and those needed to present new scientific ideas to colleagues in industry and research. I am concerned with the main writing tasks that confront professional engineers, scientists, and technical writers in their day-to-day work.

Success in fulfilling these tasks is not solely a matter of choosing a suitable style. I have suggested already the importance of the relationship between writer and reader(s), of the total context of the exchange. In judging effectiveness of communication, it is necessary to take into account not only the language chosen but also some or all the following:

- reader's familiarity with the subject;
- reader's attitude to the subject, to the writer, to the industrial organization or research centre concerned and to the journal carrying the text;
- reader's physical and mental state at the time of reading;
- reader's motivation;
- reader's expectations about style, organization and layout;
- writer's organization of material;
- writer's decisions about amount of detail and placing of emphasis;
- physical appearance of the text: layout, typesize and typeface used.

It is necessary to stress that the results of my surveys show **only** the influence of a number of linguistic features on readers' responses to a text. They show the difference that can be made by choosing language that is active rather than passive, concrete rather than abstract, and 'everyday' rather than 'special'. They show that the digestibility of statements is affected by both the length and the complexity of the sentences used to express them. They show the importance of using a variety of sentence types, some personal and some impersonal, instead of artificially restricting the choice to one or the other; and they stress the importance of breaking the whole statement (that is, the whole text) into units that give the reader the information he or she needs in convenient order and in manageable 'bites'.

The surveys do not offer evidence on **all** factors affecting readability. They simply give supporting evidence that the tactical choices recommended in this book are likely to produce an effective style suitable for specified writing tasks.

The nature of the writing task **must** be taken into account in any judgement of suitability of style. The style of a sports report, an accident report, a chemical research paper, or a lyrical description of an emotional experience must be tailored to suit three elements: the subject matter, the audience, and the context. This book does not suggest which is the best style for all writing: it suggests which is the best style for scientific and technical reports and papers, and for support documentation.

Roughly, the choices open to a writer as he or she searches the resources of language can be shown as a set of oppositions:

sentences:	short vs long
	simple vs complex
vocabulary:	short vs long
	ordinary vs grandiose
	familiar vs unfamiliar
	non-technical vs technical
	concrete vs abstract
phrasing:	normal, comfortable idiomatic expression vs special stiff scientific idioms
	direct, incisive phrasing vs roundabout, verbose phrasing
verb forms:	active vs passive
	personal vs impersonal
paragraphing:	use vs non-use
punctuation:	careful use vs casual, random use

It would be a mistake to think that good writing will be produced by consistent choice from just one side of these oppositions. Examine the best writing, and you will find variety and flexibility. Variety and flexibility are key words in our discussion. My theme is going to be that too many technical writers try to restrict their choices to formal, third-person, passive, impersonal constructions. The cumulative effect of this is a sense of monotonous, roundabout clumsiness. To produce comfortable, digestible writing, we should choose a variety of structures and vocabulary, as we do instinctively in careful conversation and description.

2.2 SENTENCE LENGTH AND COMPLEXITY

To be easy to digest, sentences must be reasonably short and not too complex. The reasons for this are not grammatical: they are connected with the number of items of information the reader can absorb in a single unit or 'thought'.

One of the basic conventions of written English is that words written as sentences, or marked off by the major punctuation marks, the colon and semi-colon, represent complete units of information or thought: the reader is to assume that the writer wants him or her to take in the information – main statement and qualifications – as single units. The number of items of information we can absorb within a unit depends on our familiarity with the subject-matter. But, however familiar we are with the subject, if there are too many items of information, we are overwhelmed. It is vital, therefore, not to pack too much information into each sentence. Consider these examples:

> According to the chemiosmotic hypothesis of oxidative and photo-synthetic phosphorylation proposed by Mitchell (refs 1–4), the link-age between electron transport and phosphorylation occurs not because of hypothetical energy-rich chemical intermediaries as in the orthodox view, but because oxido-reduction and adenosine triphos-phate (ATP) hydrolysis are each separately associated with the net translocation of a certain number of electrons in one direction and the net translocation of the same number of hydrogen atoms in the opposite direction across a relatively ion-, acid-, and base-impermeable coupling membrane (see also ref. 5).

> As a result of a critical comparison of the two standard procedures for assessing the inhibiting effect of chemicals on conidial germina-tion, namely (a) the addition of spores in aqueous suspension to a previously dried deposit of chemical, and (b) the mixing of spore suspension and chemical followed by application of aliquots to slides, it was found, using conidia of *Venturia inaequalis*. (Cke) Wint., that variability between replicates was largely eliminated by the second procedure (A.N. Author, unpub.).

> The rate of seepage of water through the sea bed into the suction anchor drains will depend on the nature of the soil and indicate conditions that where pores or fissures permit steady seepage it may be necessary to reduce the flow rate by pouring grout down pipes, to

be discharged round the anchor skirt and sucked into the pores or fissures.

However, it was soon realized that the peculiar non-experimental nature of the economic process rendered classical methods inappropriate, largely because the variables appearing on the right-hand side of the econometrician's equations are not subject to his control but are jointly determined by the simultaneous interaction of the equations in the system, so that these variables will not be independent of the error terms in the equations and hence classical least squares methods will produce biased and inconsistent estimates of structural parameters.

Readers who are not expert in the subject-matter of these examples may think that their main difficulty stems from the unfamiliarity of the vocabulary. Of course, if we do not know what the words mean, it is difficult to make reasonable judgements of the readability of texts. But even expert readers find that sentences like these leave them desperately struggling to relate all the subsidiary information to the main statement at one reading. The extracts would have been much easier to assimilate if they had been presented as several units. Even 'inexpert' readers may be surprised to find how much more manageable the statements become, even though the subject-matter remains obscure:

The orthodox explanation of the link between electron transport and phosphorylation is that it is caused by hypothetical energy-rich chemical intermediaries. Mitchell, however, explains both oxidative and photosynthetic phosphorylation with a chemiosmotic hypothesis[1-4]. He suggests that oxido-reduction and the hydrolysis of adenosine triphosphate (ATP) are each separately associated with the net translocation of electrons and hydrogen atoms: a certain number of electrons move in one direction and the same number of hydrogen atoms move in the opposite direction across a coupling membrane that is relatively impermeable to ions, acids and bases. (See also ref. 5.)

A.N. Author compared the two standard procedures for assessing how chemicals inhibit conidial germination. The procedures are:

(a) adding spores suspended in water to a dried deposit of chemical;
(b) mixing spore suspension and chemical, and then putting aliquots on slides.

He used these on conidia of *Venturia inaequalis* (Cke.) Wint. The second procedure largely eliminated variation between replicates.

The rate at which water seeps through the sea bed into the suction anchor drains depends on the type of soil. If the rate indicates that pores or fissures are allowing steady seepage, it may be necessary to reduce the seepage by grouting. Grout can be poured through pipes, discharged round the anchor skirt, and sucked into the pores or fissures.

However, it was soon realized that the peculiar non-experimental nature of the economic process makes classical methods inappropriate. The reason for this is that the variables appearing on the right-hand side of the econometrician's equations are not under his control but are jointly determined by the simultaneous interaction of the equations in the system. Accordingly, these variables will depend on the error terms in the equations, and classical least squares methods will produce biased and inconsistent estimates of structural parameters.

Compound sentence structures

A common feature of scientific and technical writing is the joining together of strings of ideas of approximately equal weight, to form long 'compound' sentences:

.....+.....+.....+.....+.....+.....+.....+.....+..... .

This stringing is usually done with the praiseworthy aim of 'getting a good flow' in the writing. Unfortunately, it often has the opposite effect: it produces unmanageable chunks that readers cannot absorb comfortably:

The degree of dependence of this pattern on bed structure and/or on production of wetted paths through the packing by randomly moving particles of the initial liquid is of interest, as in any specified packing arrangement, complete bed overload by high liquid flow-rate (pre-flooding) might be expected to result in alteration of a flow pattern, dependent on which paths through the packing were wetted, while changes in bed structure might result from re-packing or 'stirring' the bed.

In comparison, much better readability is produced by the use of comparatively simple sentences, of varying lengths, and consisting of a main statement plus at most one or two qualifications:

We thought it would be interesting to know to what extent this flow pattern depends on bed structure and to what extent it depends on the random movement of the initial liquid particles making wetted paths through the packing. In any given packing arrangement, re-packing or 'stirring' might be expected to change the structure of the bed; complete overloading of the bed with a high liquid flow-rate (pre-flooding) might be expected to alter a flow pattern, depending on which paths through the pattern were wetted.

However, it is important not to swing to the other extreme, and to produce an irritating, 'bitty' effect by producing a staccato sequence of short, simple sentences:

This flow pattern may be dependent on two factors. The first is bed structure, the second is purely random movement of the initial liquid particles producing wetted paths through the packing. It is of interest to determine to what extent flow pattern is dependent on these two factors. For a given arrangement of packing it could be expected that complete overloading of the bed by a high liquid flow-rate (pre-flooding) would alter a flow pattern. For the same arrangement, re-packing or 'stirring' the bed would be expected to cause changes in the structure of the bed.

Complex sentence structures

Another tempting way to correct a cumbersome compound structure is to re-arrange the ideas into different levels of subordination, and to express them in a 'complex' sentence structure:

. . . .
        . . . .
                . . . . .

There are three pitfalls to avoid in this 'embedding' of information in sub-clauses and sub-sub-clauses:

1. misleading the reader by faulty construction of the sentence;
2. disconcerting the reader by using not only subordinate clauses but also sub-subordinate clauses;
3. overwhelming the reader by having too many remarks in paren-thesis.

The following examples illustrate these three faults; improved versions are given for comparison.

Faulty construction

Time division multiplexed systems are basically much simpler, com-bination and separation of channels being effected by timing circuits rather than by filters and inter-channel interference is less dependent on system non-linearities, due to the fact that only one channel is using the common communication medium at any instant in time.

(Systems multiplexed by time division are basically much simpler. The channels are combined and separated by timing circuits, not by filters. Interference between channels depends less on non-linear features of the system, because only one channel is using the common communication medium at any time.)

Sub-subordination

The most important conclusion to be drawn from this analysis is that, although the signal source is not continuously connected to the load, but only for a discrete interval of time during a switching period, the actual length of time depending on the value of k, the modulated function contains the original signal information $f(t)$.

(From this analysis, the most important conclusion is this: although the signal source is not continuously connected to the load, but only for a discrete interval of time during a switching period, the modulated function contains the original signal information $f(t)$. The interval during which the signal source is connected to the load depends on the value of k.)

or

(The most important conclusion to be drawn from this analysis is that, although the signal source is not continuously connected to the load, the modulated function contains the original signal function $f(t)$. The signal source is connected to the load only for a discrete interval of time during a switching period: the actual length of time depends on the value of k.)

Excessive parenthesis

Compared with the more traditional approach of clarification by coagulation and sand filtration the MD process seems likely to yield a product of rather better appearance and having a lower content of sulphate (because of the lower usage of coagulant) and chloride (if chlorine is not used), though with a higher phosphate content.

(Compared with the more traditional approach of clarification by coagulation and sand filtration, the MD process seems likely to yield a better-looking product. The product will have a higher phosphate content but it will contain less sulphate, because less coagulant is used; and if no chlorine is used, it will contain less chloride.)

Manageability

Clearly, a key concept in producing readable writing is manageability. We must keep in mind constantly the information load that our readers will be able to accept in each sentence. Scientific and technical people often find this difficult. They are trained to be cautious in their observations and claims, so they are inclined to attach to any statement a string of conditions, qualifications or possible reservations. This frequently creates complex utterances, often with subjects well separated from main verbs. Readers just cannot hold in mind comfortably the amounts of inter-related information they are given in single statements:

If the bicarbonate ion concentration, a_1, given by the previous equation, is differentiated with respect to the hydrogen ion concentration, H,

$$\frac{da_1}{dH} = \frac{aK_1(K_1K_2 - H^2)}{(H^2 + K_1H + K_1K_2)^2}$$

the numerator set equal to zero, and the resultant equation solved for H,

$$H^2 = K_1K_2 = (4.43 \times 10^{-7})(4.69 \times 10^{-11})$$

it can be shown that the maximum bicarbonate ion concentration occurs at a pH of 8.34_1 under ideal conditions.

(It can be shown that, under ideal conditions, the maximum bicarbonate ion concentration occurs at a pH of 8.34_1. To show this, three steps are necessary:

1. to differentiate the bicarbonate ion concentration, a_1 (given by the previous equation), with respect to the hydrogen ion concentration, H:

$$\frac{da_1}{dH} = \frac{aK_1(K_1K_2 - H^2)}{(H^2 + K_1H + K_1K_2)^2};$$

2. to set the numerator equal to zero;
3. to solve the resultant equation for H:

$$H^2 = K_1K_2 = (4.43 \times 10^{-7})(4.69 \times 10^{-11}).)$$

Reversal of the membrane potential from 15 mV, the inside of the cell being negative, to around 40 mV, the inside of the cell being positive, was achieved by replacing the X with Y.

(The membrane potential was reversed by replacing the X with Y. From 15 mV with the inside of the cell negative, the potential became around 40 mV with the inside of the cell positive.)

It is important to control length and complexity, and at the same time to use a variety of structures to avoid monotony and gain emphasis. Compare the manageability and emphasis of the following two statements:

Completion of Supertron triggering is followed by commencement of anode run-down and coupling back of the negative-going waveform to the grid through V54A and a timing capacitor selected by Sw9/5, causing a 'Miller' type run down, at a speed controlled by the selected timing capacitor and RV51 setting. This run down, 'caught' at 106 volts by V52B, preventing possible jitter when the valve 'bottoms', is followed by regenerative stable state resumption by the Supertron, causing a sudden screen potential fall differentiated by C185 and R378, and triggering of the time base.

(After the Supertron has been triggered, the anode begins to run down, and the negative-going waveform is coupled back to the grid through V54A and a timing capacitor (selected by Sw9/5). This gives a 'Miller' type run-down. The speed of the run-down is controlled by the timing capacitor selected and by the setting of RV51. The run-down is 'caught' at 106 volts by V52B, preventing jitter which might otherwise occur when the valve 'bottoms'. At the end of the run-down, the Supertron regeneratively resumes its stable state; this causes a sudden fall in the screen potential, which is differentiated by C185 and R378, and triggers the time-base.)

Even short sentences can be too densely packed. The shortest poss-ible statement **may** not be the easiest for readers to digest. Consider the digestibility of the following groups of sentences:

For a given material thermal resistivity (inverse conductivity) varies linearly with electrical resistivity when either the orientation or the crystallinity vary.

(If either the orientation or the crystallinity of a material is changed, the thermal resistivity (inverse conductivity) and the electrical resistivity vary. The relationship between the resistivities is linear.)

(A change in the orientation or the crystallinity of a material changes its thermal resistivity (inverse conductivity) and its electrical resistivity. The relationship between the resistivities is linear.)

Check that with no trigger input free running does not occur at any setting of the trigger delay controls.

(Check that all settings of the trigger delay controls, there is no free running when there is no trigger input.)

(Check that there is no free running when there is no trigger input. This must be true at all settings of the trigger delay controls.)

An approximately linearly decreasing relationship of modulus of elasticity with increasing temperature resulted.

(The result was a decreasing relationship between modulus of elasticity and increasing temperature. The relationship was approximately linear.)

(This resulted in an approximately linear decreasing relationship between modulus of elasticity and increasing temperature.)

These examples make an important point: that brevity is not to be recommended for its own sake in technical writing. All three original sentences are short, but they are difficult to comprehend at first reading. The alternative versions have been 'eased out' a little, and though they use a few more words, the increases in length are amply compensated for by increases in ease of comprehension.

Clarity and readability are the features of style to strive for. Usually these are achieved by economy of words and incisiveness of phrasing; but occasionally a densely packed statement needs to be loosened to make it easier for readers to absorb.

The apparent 'density' of a piece of writing is affected also by its layout. In my surveys, many readers commented on the difference made by the punctuation and paragraphing in the various versions. The effect came not only from more careful indication of the logic and relationship of ideas by careful punctuation, but also from the improved appearance of the text when there was white space in and around the blocks of type. It is not appropriate for me to take space in this book for detailed discussions of perception, typography, and make-up of publications. Readers who would like to follow up these themes might like to look at M.A. Tinker's *Legibility of Print* [1], A.T. Turnbull and R.N. Baird's *The Graphics of Communication: typography, layout and design* [2], or M. Macdonald-Ross and E. Smith's excellent bibliography *Graphics in Text: a bibliography* [3]. Suffice it to say here that writers of reports, papers and support documentation should recognize that well-spaced type, well-paragraphed, with liberal use of subheadings, greatly increases the readability of a text.

2.3 WEIGHT AND FAMILIARITY OF VOCABULARY

It will have become obvious that there are two main features of construction that add to the inherent difficulty of the ideas we are expressing. They are:

1. the number of ideas we combine in each structural unit and the complexity with which we arrange them;
2. the weight and familiarity of the words we use to express the ideas.

The following statements are difficult to take in at first reading:

On the basis of this evidence it may be suggested that the major demand for cobalt in the nodulated legume lies in the requirement of the bacterial symbiont and that a reduction in nitrogen fixation in cobalt deficiency is in the main an indirect consequence of a reduction in bacteroid metabolism manifested by a reduced rate of proliferation...

The objective of this study was to evaluate alternative helicopter operating and support cost methodologies and approaches to assure that the degree of hardware design and logistic parameter sensitivity included in cost estimates accurately reflects actual expenditure sensitivities...

In making them easier to assimilate, it is helpful not only to break them into smaller units, but also to use more familiar terms:

This evidence suggests two conclusions: that the main demand for cobalt in the nodulated legume comes from the bacterial growth; and that a reduction in bacteroid metabolism (shown by slower reproduction) indirectly causes a lower nitrogen fixation in cobalt deficiency.

Our object was to evaluate ways of calculating the cost of operating and supporting helicopters, so that estimates of design and support costs should match actual expenditure.

Our aim should always be to write as plainly and comfortably as possible, avoiding stiff formality of expression such as:

Ruminant alkaloid toxicities associated with several herbage plants including *Phalaris tuberosa* and *Heliotropum eropeaum L.* are largely avoidable by cobalt application to the animal (A.N. Author, 1962).

In a face-to-face conversation with a visitor to the laboratory, the scientist who wrote that rather uncomfortable sentence would probably have said:

A.N. Author found that the poisonous effect of plants like *Phalaris tuberosa* and *Heliotropum eropeaum L.* can usually be stopped by giving cobalt to ruminant animals.

That alternative version would have been much easier to assimilate; it would have been just as accurate as the previous statement; it would therefore have been preferable in writing as well as in speech.

The preferred styles in all my surveys used, where possible, short words rather than long, ordinary words rather than grand, familiar rather than unfamiliar, non-technical rather than technical, and concrete rather than abstract.

A writer's first obligation is to accuracy of meaning; so where no other word or phrase will express his or her meaning accurately, he or she must use a long, technical, unfamiliar word – with supporting explanation, if necessary. But he or she should resist the temptation to use long, relatively unfamiliar technical words if there are short, familiar words available that would convey the meaning adequately. Note the important word 'adequately' in this advice. In technical literature, accuracy of meaning is paramount. Sometimes, unfamiliar technical language is essential for clear expression; in those circumstances, such language must be used; but where there is a choice, a writer should always use the short and the familiar. Why?

There are three main reasons why it is essential to choose shorter and more familiar words. One is sheer weight: shorter words have fewer syllables. They therefore help to reduce the heaviness of the writing. A second reason is the familiarity of words. Familiar words cause readers no hesitation. Each unfamiliar word causes a momentary pause as we work out its meaning. The momentary pauses have a cumulative effect, and gradually create a sense of struggle to comprehend. The third reason is inflation. If we continually choose words that are a little bigger than they need to be, eventually the whole text begins to seem inflated and pretentious.

It is not unduly fussy to look for opportunities to cut off avoidable syllables. Attention to the weight of sentences is particularly important in technical writing. It is often impossible to avoid heavy technical terms like *staphylococcus, epithelium, fertilization, ribonuclease,* and *austenitizing*. Since words like these **must** be used, it is especially important to use a minimum of syllables in building the rest of our sentences, so that the cumulative length and weight stay manageable. For example, write:

Not	*But*
alate and apterous aphids	winged and wingless aphids
hyperbaric oxygen	high-pressure oxygen
performs a function analogous to	acts like a
represents the predominant mechanism for the loss of energy	is the main way energy is lost
occupies a juxta-nuclear position	is next to the nucleus
the samples had inadequate strength properties	the samples were not strong enough

The key principle is to stay as close as possible to plain, everyday language. Write about *oil-repellency*, not *oleophobicity*; write about *rates being doubled*, not *rates being increased by a factor of two* or *being twofold enhanced*. Even apparently small changes, such as substituting *use* for *utilize*, *first* for *initial*, *kept* for *maintained*, are well worth making in sharpening the edge of a blunt and ponderous first draft.

2.4 JARGON

My advice that you should use 'plain' language wherever possible is not a condemnation of 'jargon'. Jargon is often attacked wildly as being pompous and unnecessary language, used deliberately to exclude laymen from understanding what is going on between specialists. Of course, if jargon **is** used deliberately to exclude anyone from understanding what is being said, it is indefensible. But jargon terms – special words or special turns of phrase – are not coined in order to mystify. They are created to convey between fellow-specialists precise meanings that could be expressed otherwise only at much greater length. In specialist contexts, they are the quickest and most efficient means of communication, and are wholly acceptable.

But the use of jargon has many attendant dangers. One is the temptation to use jargon outside its specialist context, to save the trouble of lengthier explanation. This almost always causes trouble for readers. Here, for example, are two statements in which the writers attached special meanings to common words. The reports in which the statements appeared were designed for readers other than specialists close to the writers:

The reaction is quantitative when the ratio of A : B is higher than 2 : 1.

The block-to-block variation must be regarded as typical rather than representative.

We cannot protest that these uses of *quantitative, typical* and *representative* are not permissible. In all spheres of activity, common words are used to express special meanings (for example, *hot* in radiation physics, *baffle* in engineering, *base* in chemistry). Our protest can be only that the writers used these words in special ways without recognizing that many of their readers would not 'understand the jargon'. In such situations, the writers were at fault. If there is any possibility that some readers will not understand a jargon word or phrase, the writer must explain it fully, or avoid using it altogether.

Another danger is that writers begin to write by stringing together jargon phrases, with an inevitable 'lumpy' effect:

Study of the **microclimatic elements** revealed that the **leaf growth rate** was limited more by **soil moisture content** than by **solar radiation effect** or **temperature influence** and that the use of the **time scales** based on **non-limiting factors** merely increased the **between-harvest variation** by producing **non-significant departures** from...

A third danger is that within a specialist context, a writer is tempted to use jargon in place of perfectly adequate everyday expressions, simply to show that he or she is a specialist among specialists. To do this is pretentious and tiring, even for readers who specialize in the same subjects as the writers.

Let me stress that this point is not made only by language specialists, protesting about their inability to understand what is going on in science and technology. Here is Sir George Pickering, Regius Professor of Medicine in the University of Oxford, protesting about unnecessary use of jargon in medical writing [4]:

> ...the function of language is to convey information accurately and from one mind to another. Let us take a few examples to show how this most elementary function of language is neglected. Here is the kind of statement that nearly every editor of a medical journal would allow: 'Unilateral nephrectomy was performed'. Most zoologists and all doctors would understand this. Would a botanist or geologist or a dentist or a physicist or a mathematician? Perhaps if he had learnt Greek; but how many now have? If he had, he would be able to translate it into his native tongue as follows: 'One-sided kidney cutting was performed'. Then why this complicated language? Why not say 'One kidney was cut out'? The information conveyed is no less precise; it is more concise; and it is understandable by all. Here is another: 'The drug induced natriuresis and kaluresis'. This is from a paper by an eminent physiologist in a first-rate physiological journal. Would a neurophysiologist understand, or the average physician, surgeon, or obstetrician? And if not, could they look up the meaning of the words? They could not. Nor, probably if they were Greek scholars, would their Greek take them all the way. I believe from the data given that this sentence means 'The drug increased the excretion, in the urine, of sodium and potassium'. This sentence, as interpreted, has the advantage that the information given is accurate and unequivocal and can be understood by all who speak English and know the elements of science. Then why did the author not say so, and why did the editors allow this sentence to pass?

I have chosen these two examples, because the authors in question are investigators of distinction, and men of culture and sensibility, and the journals that published them of the highest standing. These examples make it clear, I hope, that such writing is accepted practice, though a bad practice. Unnecessary and undesirable technical jargon is one of the bad habits of our time.

There is a considerable danger that writers who constantly use pre-packed jargon phrases will lose the capacity to find the simplest, most direct way of expressing ideas. Consider the direct statements these writers could have made:

The problem appeared to be a function of the zinc content only.
(The zinc was causing the trouble. There was no mathematical relation intended.)

...process can be applied to the determination of approximate amounts of...
(...process can be used to find out roughly how much...)

It was observed that A exceeded B by a factor of ×10.)
(There was ten times as much A as B.)

Note particularly how the casual use of the inflated expression *exceeded...by a factor of ×10* created ambiguity. The writer told me that he meant 'there was ten times as much A as B'. His colleagues pointed out that it was entirely reasonable to think that he meant there was eleven times as much.

Such unthinking use of jargon phrases leads to the danger that those phrases gradually lose the precise meaning they once carried. If *of the order of* is continually used casually to mean only 'about', we shall rapidly deprive it of its more precise mathematical meaning:

The variation is of the order of a factor of four in each batch.
(The variation is...?...in each batch.)

...accuracies of the order of ±25% can readily be achieved.
(...readily gives accuracies of about ±25%.]

...of the order of 0.5–0.75.
(?)

...of the order of three months.
(?)

...of the order of 0.2%.
(?)

...voltage drops of the order of magnitude of 10 V.
(?)

...of the order of at least tens of thousands of tons.
(?)

Gradual change in the meanings attached to words is normal and inevitable in the natural evolution of a language. It is no use complaining that it is happening. We cannot legislate to stop it. My point is that technical writers should resist the temptation to 'put on the style' a bit in their writing, to give it an air of scientific accuracy. To use words **not** because they are genuinely necessary but simply because they 'sound more scientific' is to hasten the decline in usefulness of those words. If you mean only 'approximately', or 'roughly', use one of those words. Let us preserve *of the order of* to express a precise mathematical relationship.

2.5 'FASHIONABLE' WORDS

Unfortunately, words become fashionable. They are picked up and thrown around by people who do not have a clear idea of what they mean, but who want to demonstrate that they are familiar with the latest 'buzz words'. Inevitably, these words gradually become unreliable units of exchange.

For example, the first few people who chose to use the word *functionality* in computer texts may have chosen that word deliberately to express an exact meaning. But it has been picked up and thrown around so often and so casually that ambiguous statements such as these are now common:

The X program has increased functionality.
(– works better?
– has more functions?)

To provide ABC functionality in XYZ, PSN microcode and macrocode development will be necessary.
(– to make XYZ function as though it were ABC?
– to enable XYZ to include among its functions the ability to operate in the same way as ABC?
– to enable ABC to function within XYZ?)

The word *functionality* is a legitimate English word. It is not objectionable in itself. Its use would not be objectionable if it had just one rigorously defined meaning. My point is that it is now used to mean so many things that you cannot be sure that your reader will take from it the meaning that you intended. You may have many associations in your mind when you use such a fashionable word: but before you use it, you should think very carefully whether all your readers will find the same associations surrounding that word.

Another word that has become unreliable because of fashionable over-use is *enhance* (and *enhancement*). Writers wanting to sound impressive are tempted to use *enhance* as a high-sounding synonym for *increase*. The writer of the following sentence intended to say that XYZ provides additional call-processing capabilities:

...uses XYZ to provide enhanced call-processing capabilities.

Unfortunately, that sentence could be taken to mean 'improved' capabilities: not additional capabilities, just the same range of abilities, but improved. The statement would have been more reliable if it had been:

...uses XYZ to provide additional call-processing capabilities.

To debate whether the 'correct' meaning of *enhance* is 'increase' or 'improve' is to waste your breath. It is now used commonly to mean both. And that is where the danger lies. Consider how *enhance* is used in these two sentences in a single paragraph:

Although the enhanced error rate of XYZ's lines removes the need for...will wish to make use of the enhanced speed of ABC...

When pressed to clarify this, the writer acknowledged that he meant an improved error rate, and therefore a decreased error rate, in the first sentence; and an increased speed in the second sentence. This careless use of words causes enough difficulty for native speakers of English; you can imagine, I am sure, that it causes even greater difficulty for readers using English as a foreign language.

My advice is this: if you mean 'additional', use *additional*; if you mean 'improved' or 'reduced', use *improved* or *reduced*; if you mean 'raised' or 'increased', use *raised* or *increased*. Do not be tempted to use *enhanced* because you think it has 'a better ring to it'.

Parameter is another badly abused word. *Parameter* is no longer recognized reliably to mean 'a line or quantity which determines a point, line, figure or quantity in a class of things': it is now used variously to convey the meanings 'variable', 'value', or 'limit':

There are three parameters to be taken into account.
(= variables)

Enter the parameters chosen for...
(= values)

Record the parameters of temperature...
(= upper and lower limits, boundary values)

Peripheral no longer automatically signals 'of, relating to or forming a periphery, boundary or external surface': it is now used mainly to mean 'ancillary':

...shall also require the following items of peripheral equipment...

Viable no longer principally signals a biological meaning such as 'capable of maintaining separate existence, able to live in particular circumstances': it is now used mainly to mean 'possible, workable':

...whether the scheme will be viable in present conditions...

Inhibit no longer always means 'hinder, restrain, reduce': it now frequently means 'stop':

To inhibit this, add...

When I see a statement such as *chlorpromazine inhibits rotational behaviour*, I am no longer sure whether it means that chlorpromazine 'stops' or 'restrains/reduces' rotational behaviour. I suggest it is wise, now, for technical writers to avoid using words such as *peripheral, viable,* and *inhibit* altogether.

Metaphorical expressions suffer particularly from being over-worked and casually used:

> The field of animal nutrition was the first to shed light on a cobalt requirement in nature.

H.W. Fowler's comments on the loose use of technical terms, 'popularized technicalities' he calls them, should be taken to heart. He concludes:

> ...any gratification they give to their users is at the cost of the harm done to the language by wearing down the points of words which, one suspects, may not always have been very sharp, even when confined to esoteric uses. [5]

I am not arguing that there is any intrinsic objection to the words I have discussed in this section. I am simply pointing out that many 'general' words and technical terms are now so over-worked or loosely used that you will be wise to avoid them if you can. I recognize that avoiding them will sometimes be difficult. The word *parameter* has now been embedded in many programs and on-screen displays, so you will have to use it. But if you can use *value, variable, limits* or another term in its place, it will be wise to do so.

Let me reiterate my insistence on accuracy. My call for 'plain' language is not a call for distortion, over-simplification or loss of recognizable nuances or tones. I believe, for example, that there are occasions on which it is entirely proper to use the expression *was fabricated from polyethylene sheeting;* to substitute the term *made* for *fabricated* might be to change the meaning. My point is that writers should resist the temptation to use terms like *fabricated* when *made* would serve very well, just because they feel that *fabricated* 'has a more scientific ring to it'. To do so is to run the risk of making the work sound not scientific but pompous.

2.6 'ROUNDABOUT' AND UNUSUAL PHRASING

'Roundabout' phrasing

Inflated vocabulary has an insidious effect. Soon we begin to use two or three words to express in a 'roundabout' way an idea that could have been expressed more briefly:

Samples have been subjected to examination by...
(Samples have been examined by...)

Combustion of this material can be accomplished in an atmosphere of oxygen.
(This material can be burned in oxygen.)

...is in an extreme condition of corrosion.
(...is badly corroded.)

X has a remarkable degree of stability.
(X is remarkably stable.)

...reported that the batches experienced a colour change during storage.
(...reported that the batches changed colour during storage.)

Batch-to-batch variation which occurred with X does not appear to be exhibited to the same extent by Y.
(Y does not seem to vary from batch to batch as much as X)

X is the procedure of choice from the point of view of cost.
(X is the cheapest/least expensive procedure.)

the fuel-can wall contours embrace a total thickness variation of 0.001 in.
(Fuel-can wall thickness varies by 0.001 in.)

...water stress sufficient to cause cessation of leaf elongation.
(...sufficient to stop the leaves growing longer.)

The beginning of the drought coincided with the commencement of ear emergence.
(The drought began as the ears began to emerge.)

One of the problems encountered with Glug is the difficulty experienced in extruding the material out of 2 oz tubes
(It is difficult to squeeze Glug out of 2 oz tubes.)

The storage facilities consist of a number of steel fabricated cylindrical bottles. The bottles, linked by header tubes, are sited in an area adjacent to the compressor building, the ground having been excavated for the purpose of siting the bottles, the earth being subsequently replaced such that the bottles are now subterranean.

(Storage consists of cylindrical steel bottles. The bottles, linked by header tubes, are buried beside the compressor building.)

(Storage consists of several cylindrical steel bottles. The bottles, linked by header tubes, are sited underground near the compressor building.)

Inexperienced writers may think that these rolling multi-syllabic phrases help their work to sound more scientific. In fact, most readers find such a 'roundabout' style very tiresome. In my surveys, the versions with most roundabout phrasing came consistently bottom of the poll (pp.167–216). Compare these versions of the same information:

Both stirring and pre-flooding affected the stable flow pattern, but stirring had much greater effect than pre-flooding.

Our results reveal that both pre-flooding and stirring the bed exerted a marked influence on the stable flow pattern; nonetheless, the effect of re-packing was of far greater magnitude.

It was apparent from the results that both pre-flooding and bed-stirring exerted a considerable influence on the stability of the flow pattern but that re-packing exerted a substantially higher degree of influence than pre-flooding.

As the swards were visibly different in form, the sites of the six plots in each sward were selected subjectively at first.

Selection of the six plots in each sward was based on an initial visual differentiation between the two swards.

At first, we chose the sites for the six plots in each of the two swards simply by looking at them: they were plainly different in form.

Site selection of the six plots in each sward was based on an initial visual differentiation between the sward types, whose heterogeneity in respect of morphology was macroscopically apparent.

The selection of the sites of the six plots in each sward was made on an initial visual differentiation between the two swards which owing to their different morphology were quite distinct to the naked eye.

Site selection for the six plots in each sward type was based on an initial differentiation between the two swards made visually, the

swards being quite distinct to the unaided eye due to their morphological distinctiveness.

Unusual phrasing

The normal, direct phrasing of the first statement of each of those sets
also contrasts with another common weakness of technical writing –
the use of special, stiff turns of phrase where ordinary phrasing is
available, and would be entirely adequate.

It would not seem normal if I wrote here that the readability of a text
is *vocabulary-dependent*. That would seem an awkward turn of phrase.
You would expect me to write that readability depends on the vocabulary used. But lumpy turns of phrase like *pressure-dependent* and *time-
dependent* abound in technical writing. I have even read *the receiver
amplifier is dc-current-regulated*. A more natural turn of phrase would
have been 'the receiver amplifier is regulated by direct current'. Unnatural turns of phrase such as the following are best avoided:

Speed of diffusion is rate-determining.
(Speed of diffusion determines the rate.)

...appears to be temperature-dependent.
(...appears to depend on the temperature.)

Please note that I am not suggesting that these unusual turns of
phrase are not technically accurate, or cannot be understood. My point
is simply that they seem stiffer to the reader than the normal phrasing
would have been.

Another form of unusual phrasing that is common in technical writing is the uncomfortable use of word-groups beginning with the participles *having* and *being*. It would be grammatically correct to say that I
had *bought a saloon car having two doors*; but it would be more usual for
me to use a phrasing such as *I have bought a saloon car with two doors* or
that has two doors. The use of constructions beginning with *having* in the
following examples does not make the statements inaccurate or difficult to understand; but the unusual use of *having* does contribute to
the ponderous and uncomfortable quality of the writing.

A more efficient machine having the ability to cope with...
(A more efficient machine with the ability to cope with...)
(A more efficient machine that can cope with...)

...is essentially a box having an aperture covered with aluminium
foil whose thickness is about 10 microns.
(...is essentially a box with an aperture covered with aluminium foil about
10 microns thick.)
(...is essentially a box that has an aperture...)

A request was received from a customer for a fluid having a tolerance to thermal oxidation in excess of that normally expected.
(A customer asked for a fluid with a higher tolerance to thermal oxidation than is normally expected.)
(A customer asked for a fluid that had a higher tolerance...)

Two-minute treatment of flour having an initial moisture content of 20% will give a product having all the desirable properties of flour made from steam-treated grain.
(Two-minute treatment of flour that has an initial moisture content of 20% will give a product with all the desirable properties of flour made from steam-treated grain.)

Many of these constructions with *having* seem to result from ill-organized thinking; and when that ill-organized thinking is also wrongly punctuated, readers are not only uncomfortable but also confused:

The carriage and slide can be moved by means of lead screws having coarse and fine adjustments.
(*Better*: The carriage and slide can be moved by means of lead screws, and coarse and fine adjustments can be made.)
(*Better still*: Coarse and fine adjustments of the carriage and slide can be made by means of lead screws.)

A perspex plate is provided with the machine having concentric circles engraved on it.
(*Better*: A perspex plate is provided with the machine, having concentric circles engraved on it.)
(*Better still*: A perspex plate, with concentric circles engraved on it, is provided with the machine.)

Constructions beginning with *being* are again grammatically correct but they are often clumsier than more normal constructions would have been, and are best avoided:

Write ...in these unusual conditions, which are the opposite to those in...
Not ...in these unusual conditions, being the opposite to those in...

Write ...cleanliness; the staff used plain soap to wash their hands.
Not ...cleanliness, plain soap being used for staff handwashing.

2.7 EXCESSIVE PRE-MODIFICATION

A particularly disturbing feature of technical writing is excessive 'pre-modification' – the piling up of adjectives, or words being used adjectivally, in front of a single noun:

> ...a mobile hopper fed compressed air operated grit blasting machine.

To pile up 'modifiers' in this way is utterly unnatural language behaviour. We would not normally dream of telling someone we had been to a store and bought a 'new green leather suede-lapelled patch-pocketed tie-belted jacket'. As a general rule, we recognize that listeners find it difficult to cope with the delivery of so many qualifications before the main noun. So we put some of our modifiers before it, and most of them after it. Usually, one or two adjectives, especially number adjectives and colour adjectives, come before the noun, and other modifiers come after it: 'a new green leather jacket, with suede lapels, patch pockets, and a tie belt'. It would have been much more natural – and easier to digest – if the text about the grit-blasting machine had been presented as:

> ...a mobile grit-blasting machine, fed from a hopper and operated by compressed air...

Writers who produce groups of pre-modifiers such as the string before *grit-blasting machine* are insensitive to the difficulties we have as our eyes and minds move along lines of texts, trying to assimilate the meaning of the sequence of words. In the next extract, for example, we have to try to hold in mind all the modifiers that follow *by*, while we are still waiting to find out what noun they will eventually be attached to – *roads*:

> The existing system will be extended by 5200 metres of 6 m wide sprayed tar on crushed stone roads.

(I can assure you that I did not make up that example!) That statement would be much easier to process if we were given the main noun of the group, *roads*, much sooner after *by*:

> The existing system will be extended by 5200 metres of roads 6 m wide, made of crushed stone sprayed with tar.

Frequently, too, long strings of pre-modifiers are difficult to absorb for an additional reason: not only do we have to wait a long time for the noun to which they are attached, but also we find it hard to see how all the pre-modifiers are meant to relate to one another:

A limited number of tests are made during warm-ups to obtain partial engine performance maps.
(...partial maps of complete engine performance?)
(...complete maps of part of the engine performance?)

Next time you are tempted to pile up modifiers before a noun, think of that (excruciating!) jacket I described earlier, and spread out your modifiers more manageably:

Don't write ...is deposited as discrete characteristially shaped and textured crystals.

Prefer ...is deposited as discrete crystals with characteristic shape and texture.

Don't write ...using a pressure sensitive, low temperature curing glasscloth coating varnish.

Prefer ...using a glasscloth coating varnish that is sensitive to pressure and cures at low temperature.

2.8 USE OF NOUNS AS PRE-MODIFIERS

Beware of causing noun cluster pre-modification decoding and assimilation difficulties for your readers!

In section 2.7, I discussed the discomfort caused to readers by excessive pre-modification. In that section, I warned generally about the discomfort caused by the presence of too many modifiers. In this section, I want to warn specifically about the difficulties caused by the use of nouns to pre-modify other nouns:

...is achieved by straightforward **key operation**.
(is achieved by straightforward operation of the key.)

This stage of the project will be followed by **control equipment selection and purchase**.
(This stage of the project will be followed by selection and purchase of control equipment.)

...had side-effects that necessitated **treatment discontinuation**.
(necessitated discontinuation of treatment.)

The problem will be **water loss prevention**.
(The problem will be to prevent loss of water.)

Those four examples illustrate four ways in which pre-modification with nouns often causes confusion for readers:

- confusion because pre-modifying nouns are less explicit than post-modifying prepositional constructions;
- confusion because the first noun after a preposition looks as if it is the complement of the preposition;
- confusion because a noun (other than a proper noun) immediately following a transitive verb looks as if it is the direct object of that verb;
- confusion because a noun immediately following a part of the verb *to be* looks as if it is intended to be the complement of that verb.

The following four sections discuss these causes of confusion in detail. My theme is going to be that pre-modification with nouns is grammatically legitimate and often effective, but that it is a style of writing that you should use sparingly and cautiously. Caution is necessary because pre-modification with nouns often leads to ambiguity; frequently, use of post-modifying constructions with prepositions (*of, with, by, from*) is more explicit and more natural.

As I discuss the merits of different structures, I shall have to use terms like *pre-modify*, *post-modify*, and *preposition*. Unfortunately, the terms teachers use to describe the structure of English vary, especially between the USA and Britain, and between 'traditional' grammarians and supporters of modern descriptive systems. My use of grammatical terminology is largely traditional; but for readers in the USA, and for readers in the UK whose teachers gave them no guidance on how to describe and discuss the structure of English, I shall include plenty of extracts from technical texts, to make clear what I mean by the grammatical terms I use.

Lack of explicitness of pre-modifying nouns

Pre-modifying nouns are frequently less explicit than post-modifying prepositional constructions. For example, when I first read the expression *is achieved by straightforward key operation*, I had to pause to work out its meaning. My first instinct was to read the words *straightforward* and *key* as a linked pair – 'a straightforward key'; but that interpretation was obviously absurd, so I had to adjust my interpretation to 'straightforward operation of the key'. Of course, that adjustment took only a fraction of a second; but if the writer had used the post-modifying *of...* construction in his or her original text, the meaning would have been explicit immediately.

Similarly, in the example, *can be used for direct user input*, confusion arises because in English, a reader's first expectation is that, when a noun is used to pre-modify another noun, the pre-modifying noun is intended to have a meaning equivalent to a post-modifying *of...* construction:

pre-modifying noun	main noun		equivalent post-modifying *of...construction*
information	retrieval	=	retrieval of information
knowledge	representation	=	representation of knowledge
data	collection	=	collection of data
speed	reduction	=	reduction of speed
time	management	=	management of time

Accordingly, we hesitate (albeit only briefly) when our eyes and minds come across word-groups such as:

...can be **used for direct user input**.
...the only system that **provides user feedback**.
...**facilitates operator adjustments**.
...arrangements **for customer reception**.

The meaning of the bolded groups is not immediately explicit. When we read *for direct user input*, our first response is to assume that it means 'direct input of users'; but we recognize at once how absurd that would be, so we assume the writer wants us to supply a different interpretation: 'direct input by users' or 'direct input from users'.

The mental adjustment we have to make in decoding that word-group takes only a fraction of a second; but the writer would have removed the need for even that small adjustment if he or she had written explicitly *the pen can be used for direct input by users*.

Similarly, it takes us only a moment to realise that *provides user feedback* and *facilitates operator adjustments* cannot mean 'provides feedback of users' or 'facilitates adjustments of the operator'; but we would have been saved even these brief moments of adjustment if the writers had written explicitly *provides feedback from users* and *facilitates adjustments by the operator*.

The word-group *arrangements for customer reception* presents particular difficulty. In the text in which I found that group, it could have meant:

arrangements for reception of customers
arrangements for reception by customers
arrangements for reception from customers

The ambiguity would have been removed if the writer had used an explicit post-modifying construction, with the preposition he or she intended – *of*, *by*, or *from*.

In general, when the noun that is pre-modified expresses an activity (physical or mental), the pre-modified version usually expresses a meaning equivalent to a post-modifying *of...* construction:

output reduction	=	reduction of output
machine maintenance	=	maintenance of machines
data entry	=	entry of data
case study	=	study of a case
patient hypnosis	=	hypnosis of a patient
rock disintegration	=	disintegration of rock
machine minding	=	minding of machines
data processing	=	processing of data
problem solving	=	solving of problems
flight scheduling	=	scheduling of flights
wire cutting	=	cutting of wires

If this is so, can we set up a guideline that says 'Pre-modification with words that normally function as nouns **always** carries a meaning equivalent to a post-modifying *of...* construction'?

Unfortunately, we cannot. In English, it is entirely grammatical, and quite common, for us to use expressions such as:

Group 1

a student nurse	(which does not mean a nurse of students)
a steel tray	(which does not mean a tray of, or loaded with, steel)
a radio beacon	(which does not mean a beacon made of radios)
Herschel fringes	(which does not mean a fringe consisting of Herschels)
a glass container	(which does not mean a container of glasses)

Group 2

a computer simulation	(which does not mean a simulation of a computer)
a wheel arch	(which does not mean an arch of wheels)
an acid pipeline	(which does not mean a pipeline of acid)
the laboratory values	(which does not mean the values of laboratories)
night supervision	(which does not mean supervision of the night)

In grammatical terms, these pre-modifying nouns all act adjectivally (or attributively), but they act in different ways in the two groups. The pre-modifying nouns in the first group indicate intrinsic attributes of the nouns they qualify; they act simply as adjectives:

a student nurse	=	a nurse who is a student
a steel tray	=	a tray made of steel
a radio beacon	=	a beacon that consists of (or transmits) radio signals
a glass container	=	a container made of glass
Herschel fringes	=	fringes (in a thin film) that have the qualities first described by Herschel in 1809

The pre-modifying nouns in the second group act attributively in that they express a functional relation to the nouns they qualify. To express their meaning in another way, we should have to use a post-modifying construction with *by*, *for* or another preposition:

a computer simulation	=	a simulation by (means of) a computer
a wheel arch	=	an arch for (or to accommodate) the wheel
an acid pipeline	=	a pipeline for acid
laboratory values	=	values found in laboratory work
night supervision	=	supervision at night

As these examples show, pre-modifying nouns can be equivalent to simple adjectives, or to post-modifying constructions with prepositions other than *of*. We cannot set up a guideline that, says **all** pre-modifying nouns are equivalent to post-modifying *of...* constructions.

The implication of this is that we must think carefully about what we want to imply by use of a pre-modifying noun. In general, I urge you to use nouns as pre-modifiers sparingly: prefer the greater explicitness of post-modification. In making this recommendation, I am offering stylistic advice, not grammatical advice; but it is stylistic advice supported by evidence and judgements from eminent grammarians. In *A Comprehensive Grammar of the English Language* [6], probably the most thorough analysis of English ever made, Randolph Quirk and his colleagues comment:

> In general, premodification is to be interpreted (and, most frequently, can only be interpreted) in terms of postmodification and its greater explicitness.

They go on to acknowledge, however, that when word-groups containing pre-modifying nouns are used frequently in a given context, readers often are not troubled by possible ambiguity: they select immediately a meaning appropriate to the context.

If you frequently read texts related to computing or electronics, I am sure you are familiar with expressions such as *computer simulation* or *laser scanning*. Perhaps, when you first saw them, you interpreted those expressions to mean 'simulation of computers' and 'scanning of lasers'; but not for long. You soon learned to jump immediately to the 'adjusted' meanings 'simulation by (or in) computers' and 'scanning by (or with) lasers'. But note what you had to do: you had to guess what the pre-modified group **probably** meant. Fairly certainly, you made the right guess; but if the writers had used the post-modifying prepositional structures instead of the potentially ambiguous pre-modifications, you would not have had to guess the meanings.

My recommendation is based on a 'safety-first' attitude: take every opportunity to remove the need for readers to guess what you probably mean. I make this recommendation heartily because when I am reading technical texts, I am so often bewildered by careless use of pre-modification. As a final example for this section, here is an apparently innocuous statement:

> ...can be configured to meet a wide range of **user data communication requirements**.

As I thought about what that could mean, I could see six possible interpretations:

- to meet a wide range of requirements for communicating users' data...
- to meet the data-communication requirements of a wide range of users...
- to meet the requirements for communicating data from a wide range of users...
- to meet the requirements for communicating data to a wide range of users...
- to meet the requirements for communicating a wide range of data about users...
- to meet the requirements for communicating data about a wide range of users...

Did the writer intend to talk about 'a wide range of users', 'a wide range of data', or 'a wide range of requirements'? Did he or she intend to talk about 'users' data' or about 'users' requirements'? And did he or she intend to talk about 'users' data' (data belonging to users), about 'data from users' (received from them, though not necessarily their data), about 'data in general' (to be communicated to users), or about 'data about users'?

The intended meaning of the four-word group *user data communication requirements* was masked by the writer's careless use of a string of three nouns to pre-modify *requirements*. Once again, we can see that noun pre-modification (better: pre-modification with nouns!) is often a dangerous practice.

So, I recommend that you use post-modifying prepositional constructions to avoid many of the problems of explicitness/ambiguity that are caused by pre-modifying nouns. But, from discussions with dozens of writers who have been tempted to use pre-modifying nouns, I know that they often use that style because they are afraid of creating extremely clumsy sequences of post-modifying prepositional constructions.

I acknowledge readily that the stringing together of **many** prepositional constructions is likely to create a cumbersome effect. For example, it would not have been desirable to revise the statement about *user data communication requirements* in this clumsy way:

...can be configured to meet a wide range **of the requirements of users for communication of data**.

Certainly, this would have been explicit, and would have been preferable to the use of the four pre-modifying nouns; but it would have been uncomfortable to read, and therefore not good style.

We need to ask ourselves whose requirements we are talking about, and what sort(s) of requirements they are:

Whose requirements?	The requirements of users.
What sort(s) of requirements?	Requirements for communicating data.

A step towards a more comfortable and natural way to express that information would be to write:

...can be configured to meet a wide range of users' data-communication requirements.

But there is still ambiguity in that version. Although we have made clear that we are talking about users' data-communication requirements, it is still uncler whether *a wide range of* relates to *users* or to *requirements*. The best version, entirely explicit, would therefore be to use one pre-modifier (*data-communication*) before *requirements*, and to use just two post-modifying groups afterwards:

...can be configured to meet the data-communication requirements of a wide range of users.

Confusion between pre-modifying nouns and the complements of prepositions

A second way in which pre-modifying nouns cause confusion is that they often seem to be the complement of a preposition.

Prepositions are words that introduce (or govern) nouns, noun phrases or clauses (such as *wh-* or *-ing* clauses):

under the lid
for introduction of the cannula
from what we have learned previously
through which the gas passes
before opening the door
by determining the frequency.

When the complement of a preposition is a noun or noun phrase, it normally appears as the first noun after the preposition:

...is used by **operators** to create
...is fixed above **the door**.

Usually, articles (*the, a, an*) and adjectives can come between a preposition and its complement without confusion:

...is used by **skilled operators** to create
...is fixed above **the low, green door**.

But when a noun is used as a pre-modifier, confusion often arises, because readers assume – at least momentarily – that the pre-modifying noun is the complement of the preposition:

...two items of equipment are needed for the **power** subsystem maintenance
...provide a high degree of **noise** immunity.

In both those examples, by using post-modifying prepositional constructions, the writers would have saved readers the trouble of having to adjust their interpretation:

...two items of equipment are needed for **maintenance** of the power subsystem
...provide a high degree of **immunity** from noise.

When the true complement gets pushed even further from its preposition, the greater explicitness and the greater ease of reading of a post-modifying construction become even clearer.

Compare:

...this stage of the project will be followed by control equipment selection and purchase

with:

...this stage of the project will be followed by selection and purchase of control equipment.

Compare:

...energy is absorbed in the process of resistor material vaporization

with:

...energy is absorbed in the process of vaporization of resistor material
(or **better**: energy is absorbed as the resistor material vaporizes).

Consider, too, the confusion caused by the following statement, in which the pre-modifying noun creeps between the preposition and its true complement, **and** the use of a pre-modifying noun leaves the reader unsure of whether the intended meaning was 'control **of** the XYZ system' or 'control **by** the XYZ system':

pressure on the ON LINE pushbutton causes the ABC to go on-line, readying it for XYZ system control.

(The intended meaning was 'control **by** the XYZ system'.)

Once again, I urge you to avoid using nouns as pre-modifiers. Prefer the more explicit use of post-modifying constructions:

Don't write	Three busbars are supplied for purposes of failure isolation.
Prefer	Three busbars are supplied for the purpose of isolating failures.
or Better	Three busbars are supplied for isolating failures.
Don't write	. . .have difficulties with the task of form completion.
Prefer	. . .have difficulties in completing forms
Don't write	Devise appropriate methods of answer representation.
Prefer	Devise appropriate methods of representing answers.

Once again, in making this recommendation, I am offering stylistic advice, not grammatical advice; but it is advice based on a recognition of how we read, and how we can make text as easy as possible to comprehend.

As our eyes move along the lines of a text, our minds attempt to 'close' round recognizable word-groups, and to pass them into our short-term memories for assimilation into the statement that is being assembled there.

We arrive at the expression:

'The equipment and materials needed for the main landing gear. . .'

The word-group *for the main landing gear* looks like a familiar, plausible, complete unit, so we absorb that as a chunk of meaning, and move on. We take in the next few words:

'. . .strut servicing are. . .'

We realize that we should have related the preposition *for* to the word *servicing*; so we rapidly re-arrange the meaning we have formulated, and obtain:

'The equipment and materials needed for servicing the main landing-gear strut are. . .'

Of course, our minds do that in a fraction of the time it has taken me to discuss what happens as we read, but you can see that once again, the writer's choice of a pre-modified style has caused us an unneces-

sary moment of adjustment. If he or she had placed the true complement *servicing* close to its preposition *for*, we should have been able to
move unhesitatingly along the line of words, concentrating on absorbing the statement, without the distraction of having to disentangle the
phrasing.

Confusion between a pre-modifying noun and the direct object of a transitive verb

A third way in which pre-modifying nouns cause confusion is that they
often look as if they are the direct object of a transitive verb. The first
time I read the following statement, I was momentarily confused:

Drug X had side-effects that necessitated treatment discontinuation.

I re-read the sentence twice, to ensure that I had not misunderstood
the statement.

My confusion was due to the placing of the noun *treatment* immediately after the verb *necessitated*. The verb *to necessitate* is what
grammarians describe as a 'transitive' verb: it needs to be completed by
mention of an 'object', a person or thing: the drug had side-effects that
necessitated...(something). That something seemed to be *treatment*.
But then the word *treatment* was followed by *discontinuation*. I was
obliged to adjust the meaning I had passed back into my short-term
memory. The sense of the statement had to be completely reversed:
I had to think not about the necessity of *treatment*, but about the
necessity of *discontinuation* of treatment. I had been misled by the use
of *treatment* as a pre-modifier. When I converted *treatment discontinuation* into the post-modified group *discontinuation of treatment*, all was
clear.

Once again, I have to acknowledge that my mind made the adjustment in a fraction of the time it has taken me to discuss the point here;
but if the writer had placed *discontinuation* immediately after
necessitated, to give *necessitated discontinuation of treatment*, I should have
been able to concentrate entirely on assimilating the meaning of the
statement; I should not have been disturbed even briefly by having to
disentangle the phrasing.

I acknowledge, too, that the occasional need to make adjustments
such as I have been discussing does not usually have a disastrous
impact on a reader's ability to concentrate on the incoming message;
but if pre-modification is a frequent feature of a writer's style, its
cumulative effect is to make the text seem lumpy and uncomfortable to
read.

So look for every opportunity to remove potential sources of disturbance:

Don't write	...are added to 'commercially pure' titanium to improve its corrosion resistance.
Prefer	...are added to 'commercially pure' titanium to improve its resistance to corrosion.
Don't write	Flow meters ensure accurate material and catalyst measurement.
Prefer	Flow meters ensure accurate measurement of materials and catalysts.
Don't write	Aspirin brings pain relief
Prefer	Aspirin relieves pain

Confusion between pre-modifying nouns and the complement of verbs like *to be* or *to become*

The fourth way in which pre-modifying nouns cause confusion is that they often look as if they are the complement of an 'incomplete' verb like *to be* or *to become*.

The verbs *to be* and *to become* have to be completed in some way. We cannot say simply 'The lamp was.', 'The beam became.', 'The container is.', or 'The answer is.'. We need to provide a completion – in grammatical terms, a *complement*:

The lamp was broken.
The beam became brittle.
The container is a rectangular box.
The answer is to increase X.

As these examples show, we can complete the verbs *to be* and *to become* in many ways – with adjectives (*broken*, *brittle*), noun phrases (*a rectangular box*), and infinitive constructions (*to increase X*). Unfortunately, when a noun complement is pre-modified with another noun, readers are often momentarily confused. For example, I stopped momentarily when I read this expression:

...the problem will be water loss prevention.

Conventionally, the complement of the verb *to be* is kept as close as possible to the verb itself. So, since *water* followed *will be*, my mind began to form the idea of *water* being the problem. But then I read the remaining words, and I had to adjust my interpretation. I recognized that the intended complement of *will be* was *prevention*. It was therefore necessary for me to adjust the meaning in my mind before I could go on to assimilate the rest of the writer's argument.

Yet again, the disturbance to my concentration was only momentary; but there **was** a disturbance, caused by the unnatural, pre-modified group *water loss prevention*. If the writer had put *prevention* immediately after *will be*, the meaning would have been clear. Indeed, if he or she had adopted a more direct style, avoiding so many noun forms, the text could have been much more natural:

...the problem will be to prevent loss of water.

Wherever you can, avoid use of pre-modifying nouns after *to be* and *to become*:

Don't write	The outcome was sludge and water deposition.
Prefer	The outcome was deposition of sludge and water.
Don't write	This species of tree is lime tolerant.
Prefer	This species of tree tolerates lime.

2.9 ABSTRACTION

One reason for excessive indulgence in roundabout phrasing and pre-modifying nouns is technical writers' taste for abstractions. Some writers always seem to express themselves through abstract nouns rather than active verbs:

> The thermal **decomposition** of the TMAH **occurs** rapidly at these temperatures and **results in the formation** of trimethylamine and methyl alcohol.

> ...**brings about** the characteristic response of **emergence**.

> ...**causes stimulation** of yield by the application of fertilizer to...

> **Contraction** of the tree stems **occurred** rapidly.

In all these examples, an abstract noun and a roundabout phrase are used where active verbs could give directness and life to the statements. These direct, active versions have much more vigour and momentum:

> The TMAH decomposes rapidly at these temperatures and forms trimethylamine and methyl alcohol.

> ...usually makes them emerge/come out.

> ...stimulates the yield by putting fertilizer on...

> The tree stems contracted rapidly.

Too much indulgence in abstractions is almost as blurring as too much use of jargon. As Robert Gunning observes:

> It is difficult to pin down terms like 'capital disequilibrium' or 'idealistic parallelism'. Such terms are very far up the ladder of abstraction. And the farther up that ladder a word is, the less chance it has of meaning to the reader just what it did to the writer. [7]

By a ladder of abstraction, Gunning means a series like:

energy source
fuel
liquid fuel
gasoline
hexane

The bottom one of these is the most specific and concrete: the top is the most general and abstract. Each term has a proper place in our range of vocabulary; but that place depends on what we are trying to say. There are times when we want to generalize about a 'vehicle' or a 'dwelling' rather than to specify a particular type of car or house. In those circumstances, the abstraction is the best word to choose. But the writer who wrote *we have lacked personnel capacity* made me stop and wonder if he meant more than 'we have been short of staff'. I had to puzzle over the statement *the incidence of each of these items differs with different causal factors*; subsequently, I discovered that it meant simply 'each item has a different cause'. And I lost track completely because of the switch from *paint* to *system* in the statement *the paints were examined again and it was found that the viscosity of all systems had increased.*

For effective writing, the principle is: do not go up the ladder of abstraction further than the sense requires. Do not write *fuel* when you mean specifically *hexane*; do not write *paint* at the beginning of a sentence, and change to *system* later in the same sentence. Take note of Sir Ernest Gowers' advice about the temptation to use abstractions:

> Unfortunately the very vagueness of abstract words is one of the reasons for their popularity. To express one's thoughts accurately is hard work, and to be precise is sometimes dangerous. We are tempted to prefer the safer obscurity of the abstract. It is the greatest vice of present-day writing. Writers seem to find it more natural to say 'Was this the realization of an anticipated liability?' than 'Did you expect to have to do this?': to say 'Communities where anonymity in personal relationships prevails' than 'Communities where people do not know one another'. To resist this temptation, and to resolve to make your meaning plain to your reader even at the cost of some trouble to yourself, is more important than any other single thing if you would convert a flabby style into a crisp one. [8]

Notice the contrast between the 'crisp' style and the 'flabby' style, between the active constructions and the abstract circumlocutions, in the following pairs:

Measurement of the torque is achieved by means of the plastograph.
(The plastograph measures the torque.)

To enable technicians to pursue printed circuit board servicing efficiently...
(To enable technicians to service printed circuit boards efficiently...)

Notice, too, that it is not only nouns that can be arranged in ladders or levels of abstraction. You have probably seen before a typical 'tree structure', emphasizing the way in which some nouns name classes that cover/include a number of other classes 'below' them; that is, how some nouns are higher-order generalizations than others:

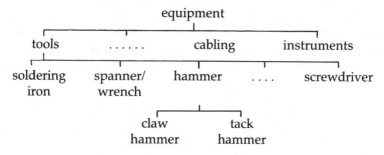

But you may not have recognized previously that verbs and adverbs also can be arranged in levels of abstraction:

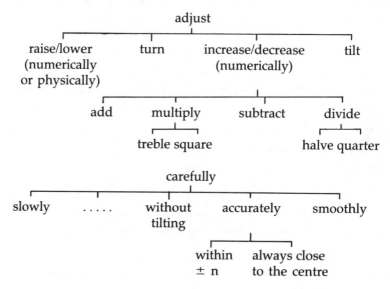

Higher-order abstractions are valuable: they help us generalize. For example, sometimes we want to refer generally to the *behaviour* of an aircraft wing or of a structural beam; we do not want to refer to a specific concept like *vibration* or *deformation*; so we choose a higher-level abstraction intentionally. This is desirable vagueness or generalization. **But** the higher the level of abstraction, the greater the chance of unintended vagueness creeping in. So, use the lowest level of abstraction at all times – be as specific as possible:

Don't write	*Prefer*
shall make use of a lifting facility	shall use a crane
shall carry out measurement of certain parameters	shall measure weight and temperature
aircraft has high-altitude operating capability	aircraft can fly above 60 000 feet
will be three vertical pedestrian access points	will be three lifts (elevators)

2.10 EXCESSIVE 'NOMINALIZATION'

When we are tempted to express ourselves in a more abstract way, we are tempted into excessive 'nominalization' – excessive use of 'noun-centred' structures.

From time to time, you will want to focus on an abstract concept: for example, on the notion *allocation*. In such circumstances, you will often need to use a passive, roundabout construction to express your thought. For example:

...fair **allocation** can be achieved only by...

But beware of falling into the habit of expressing all your thoughts in constructions that consist of abstract nouns plus colourless 'general purpose' verbs (like *achieve, accomplish, perform*). If you do, you will often move the focus of your statement unintentionally. For example, note how the focus of each of these 'noun-centred' statements falls on the *functions* that are performed:

The functions of allocating and apportioning revenue are performed by the ABC.

The ABC performs the functions of allocating and apportioning revenue.

The writer wanted to focus on the activities of *allocating* and *apportioning*. A crisper, 'verb-centred' style, focusing on the activities, would have been:

The ABC allocates and apportions the revenue.

If you find yourself using an abstraction plus to *take place, occur, perform, effect, achieve, accomplish, result, carry out, conduct, observe, find, be seen,* look for a more direct way of making your statement, usually by converting the abstract noun into a verb:

Write X corrects errors by retransmitting A and B.
  Not X carries out error correction by retransmission of A and B.
Write The time required can be reduced.
  Not A reduction in the time required may be effected.
Write This offers the possibility of correlating X with Y.
  Not This offers the possibility of performing a correlation of X with Y.

Write   To integrate the signal, a storage device is needed.

Not   To effect the signal integration, a storage device is needed.

Write   Such a record can be retrieved only sequentially.

Not   Retrieval of such a record can only be done in a sequential manner.

Write   Routing modules also control congestion and the lifetime of packets.

Not   Routing modules also provide congestion control and packet lifetime control.

Write   ABCD Department are at present measuring the flow properties of X.

Not   ABCD Department are at present carrying out flow property measurements on X.

Write   This method of checking fuel flow...

Not   This method of carrying out fuel flow checks...

2.11 VERBS: TENSE AND VOICE

Young engineers and scientists are not to be blamed for excessive use of abstract roundabout style, because usually they are simply trying to observe a 'rule' imposed on them by less discerning teachers of science and technology than Sir George Pickering (whose advice on abstract jargon was quoted on page 22). The 'rule' is that technical reports and papers must be written in the past tense, passive voice and third person (or impersonal). It is a bad rule. I believe it is responsible for much of the stupefying ponderousness of expression in technical writing.

Tense

First, tense. It is rarely possible to write a technical document, even a scientific report, **wholly** in the past tense. Certainly, large parts of a report should be in the past tense. It is natural to use past tenses to state what the objectives were, what equipment was used and what procedure was followed. But from the very first sections, it is frequently necessary to talk about continuing objectives or persistent conditions, using present tenses; in descriptions of procedures, it is often necessary to quote general principles or 'eternal truths', using *present* tenses; and in discussions of data or results, it is usually necessary to state what the results *show* not *showed* and what you *deduce*. Here are some examples.

Natural mixture of tenses in introductions

Routine levelling on the site revealed certain discrepancies in the levels indicated on the brass-plate bench-marks. There are three types of bench-marks on site: ordnance survey bench-marks stamped on walls; brass plates mounted on concrete plinths and stamped with both the height in feet above Newlyn datum and the level in feet with respect to site datum; and angle plates mounted on concrete plinths. Our brief was to review the site bench-mark system, taking the ordnance survey bench-mark at Admin Block as the site reference level, and to check all other marks with respect to this level...

The estimator engaged in concrete construction must be aware of the cost implications of a particular specification in order to draw up enquiries for ready-mixed concrete and analyse the subsequent quotations. This knowledge is particularly important when the use of site-mixed concrete is envisaged. The experiments described in this paper were designed to...

Natural mixture of tenses in descriptions of experimental procedures

The samples were examined by means of microscopical surface views (both sides), microscopical views of the separated warps and shutes, and cross-sectional micrographs. To obtain the sectional view, the wire must be embedded in plastic, then ground and polished. This gives a profile such as is shown in Figure 1. To minimize the influence of the plastic coating, the material used was...

The horizontal diameters of both the six-segment ring of ordinary grey iron and the twelve-segment ring of SG iron were measured twice: immediately after they were built and while they were in the tail of the shield. The next measurements, taken five days later, showed the horizontal diameters of both rings had decreased. The deflection was of the opposite sense to the distortion of tunnel linings usually observed in clay. Normally there is an increase in horizontal diameter and a decrease in the vertical diameter, commonly known as squatting...

Natural mixture of tenses in discussions of results

The relations obtained between dry density and number of passes with the roller show that, for the moisture contents and depth of layer used, compaction increased only slightly in the cohesive soils and in the uniformly graded fine sand after four passes of the machine (Figure 3). The compaction increased considerably when the number of passes was increased from four to eight. An air content of 10% or less is considered to be satisfactory when...

Part of the H_2 fraction was not retained on QAE-Sephadex at pH 8.3. Analysis of this fraction by isoelectric focusing and SDSP gel electrophoresis showed it was heterogeneous, but contained predominantly low-molecular-weight species of protein, ..., ..., and ...... These species are rich in glycine, serine, and glutamic acid, and contain relatively low amounts of basic amino acids. Therefore, it was assumed that the basic nature of these unretained proteins resulted from extensive amidation of the free carboxyl groups...

In the conclusions section of a paper, it is natural to make *present-tense* statements about the position that has been reached. In the recommendations, it is essential to state what should be done in the future. It just is not possible to write accurately, yet at the same time to observe an over-all rule that says reports must be wholly in the past tense.

Voice

It **is** possible to write papers almost entirely in the passive voice; but it is usually distorting and tortuous to do so. You run the risk of unintentionally changing your emphasis and meaning, and of increasing the verbosity of your text.

'Voice' is a grammatical term used to describe the possibility we have in English of viewing the action of a sentence in two ways without changing the facts reported:

John designed the laboratory (active).
The laboratory was designed by John (passive).

Acid-treatment removed the rust (active).
The rust was removed by acid-treatment (passive).

The subject position in a sentence is where we normally place (and look for) the topic of that sentence. The active sentences above make statements about *John* and *acid-treatment*: the passive sentences make statements about the *laboratory* and *rust*. By using the passive construction, we move the centre of interest from the 'performer' to the 'undergoer' of the action. Though the information remains essentially the same, the emphasis is different.

The passive voice is a valuable feature of the English language, and one widely used in all forms of writing and speaking, not just in science and technology. But it is vital to note that the active and passive voices are used for specific purposes, to create deliberate balance or emphasis in a statement. They should not be interchanged arbitrarily or haphazardly.

The usual way of arranging a statement in English is in two parts: a topic and a comment.

Topic	Comment
1. The CAN command	cancels data received after the last terminator.
2. The junior technician	added extra solvent to the flask.
3. The surgeon	will open the artery at two points.
4. The road repair gang	has broken the water pipe in three places.

5. My new doctor	believes that aspirin damages the stomach.
6. The Managing Director	will announce on Friday that he will sack 500 people on 31st December.

In the topic, we announce what we want to talk about; in the comment we present the news – what we want to say about the topic.

Normally, as in the six sentences above, the verb is ACTIVE. In our comments, we talk about what the subject of the sentence **is** or **does** (1 and 5, present tenses), **was/were** or **did** (2 and 4, past tenses), or **will be** or **will do** (3 and 6, future tenses). But sometimes we want to talk about a topic in which the subject does not **do** something. On the contrary, our topic has something **done to** it or **thought about** it. In those circumstances, we use the PASSIVE form of the verb:

Topic	*Comment*
1. Data received after the last terminator	are cancelled by the CAN command.
2. Extra solvent	was added to the flask by the junior technician.
3. The artery	will be opened by the surgeon at two points.
4. The water pipe	has been broken by the repair gang in three places.
5. Aspirin	is believed by my new doctor to damage the stomach.
6. An announcement	will be made by the Managing Director on Friday that 500 people will be sacked on 31st December.

As before, the topic announces what we want to talk about; the comment presents the news – what we want to say about the topic.

In general, we form a passive structure by moving the 'agent' to the end of the sentence, usually preceded by *by*. But we also use passive structures without reference to any agent, in specific circumstances:

- when we think the information about the agent is obvious or unimportant:
  1. Data received after the last terminator are cancelled.
  2. Extra solvent was added to the flask.
  3. The artery will be opened at two points.
- when we do not know the identity of the agent:
  4. The water pipe has been broken in three places.
- when we want to assert a generally held belief:
  5. Aspirin is believed to damage the stomach.

- when we do not want to state who is/was responsible for an idea/
  action:
  6. It will be announced on Friday that 500 people will be sacked on
     31st December.

So we must think hard about what we want to say. To make an
arbitrary change from *Two thin struts linked the plates to the rig* to *The
plates were linked to the rig by two thin struts* is to change the emphasis
from the *struts* to the *plates*. The following two statements are not exact
equivalents. The choice of an active version or a passive version should
not be dictated by an arbitrary rule of style:

Absolutely parallel loading was produced by adjustment of the
Schultz plates.

Adjustment of the Schultz plates produced absolutely parallel
loading.

Skilful writers increase the precision with which they convey meaning by deliberately moving from active to passive. Unskilled writers,
moving from active to passive simply to satisfy a dimly understood
'rule', frequently decrease the precision with which they put their
thoughts into words.

Shift from active to passive can do more than just change emphasis
or balance: sometimes it changes meaning. This happens when the
verb that is converted from active to passive contains one of the
auxiliary verbs such as *can, will, would,* or *might*.

In an active sentence, *can* is usually taken to mean 'is able to'; in a
passive sentence, it is usually taken to mean 'is possible'. Consequently, to change *The wires can support the sheet* to *The sheet can be supported by
the wires* is to signal a different meaning to the reader. *A three-port
matrix can represent the whole transformation* should not be lightly inter-
changed with *The whole transformation can be represented as a three port
matrix*. *Thermostatic valves can regulate the room circuit* does not necessarily signal the same information as *The room circuit can be regulated by
thermostatic valves*. Active and passive constructions should not be arbitrarily inter-changed: blanket imposition of passive forms may force
writers into inaccuracy.

Not only does the passive enable us to move the performer or agent
to a subordinate position, and give more stress to a more important
object or activity; it also enables us to omit the agent altogether when
the agent is not relevant or not known. It is this ability that makes the
passive construction attractive in technical writing. Technical writers,
especially report-writers, normally wish to emphasize the procedures
and findings of their work, not the people who did it. They want

accounts of their work to be as detached or impersonal as possible. Use of passive constructions achieves this.

In descriptions of experimental procedures, it would usually be irrelevant and unnecessary to state precisely who lit a burner or soldered a joint, so we do not write:

The technician in our laboratory reacted A with B.
My assistants and I connected the pin to the shank.

We write, quite naturally:

A was reacted with B.
The pin was connected to the shank.
When the painted surfaces were examined last September...
Data from the radar equipment were processed rapidly.
Output from the reactor has been increased.
Reed relays were fitted to the board, which was installed...

Technical writers are not unusual in using passive constructions in this way. Four out of five English passive sentences have no agent openly expressed [9]. We cannot object to this use of the passive construction in itself. We **can** object to its **abuse** – to use almost to the exclusion of all other constructions. When the passive is used as a rule, not as an exception to obtain a particular effect, writing soon begins to seem forced and uncomfortable.

Though the proportions of active and passive constructions used by individuals vary considerably, the active voice is the predominant mode of expression for native English speakers. If it were not 'normal' for English statements to appear in the word order:

Subject	+	Verb	+	Object
(actor or agent)		(action or state)		(person or thing affected)

it would not be possible for us to gain particular effects by deliberately changing the word order or by changing the items used to fill the subject or object positions. To try to use only passive constructions is, therefore, bound to produce writing that seems artificial and forced; and as we have seen, it is likely to lead to distortion of meaning.

Since passive constructions are formed by adding parts of the verb *to be* (or sometimes *to get* or *to become*), plus *by* if the agent is expressed, continual use of passives commonly increases roundabout phrasing and length.

Increase in length is not inevitable. *We calculated the yield* can become *The yield was calculated*; but it frequently expands to *Calculation of the yield was performed*. Especially where an agent must be specified, extra words slip in. *The ohmmeter measures the resistance* becomes *The resistance is measured by the ohmmeter* or *Measurement of the resistance is carried out by means of an ohmmeter*. I shall comment in the next section (p. 65) on the introduction of 'general-purpose verbs' such as *carried out*. For the moment, let me just draw attention to the greater length of the first (mainly passive) versions in the following pairs:

> The first test was carried out on the tower aerials in order to determine the vertical angle response. A helicopter with a transmitter suspended beneath it flew to various points around the tower while radar equipment was used to track it, phases were measured by means of Measurit A and XY-5 performed calculations of the azimuth angle of the helicopter. The helicopter flew to six points spaced at equal intervals on a circle of radius about three miles centred on the tower. This procedure was carried out five times at heights of 1000, 2000, 3000, 4000 and 5000 feet above ground level, the helicopter hovering for about 30 seconds at each point, and transmissions being produced for 15 seconds at 5.685 MHz and for 15 seconds at 8.975 MHz.
>
> (In the first test to establish the vertical angle response of the tower aerials, a helicopter with a transmitter suspended beneath it flew to various points around the tower. Radar equipment tracked the helicopter, Measurit A measured the phases, and XY-5 calculated the azimuth angle of the helicopter. The helicopter flew to six points equally spaced on a circle three miles in radius centred on the tower. It hovered for about 30 seconds at each point, transmitting for 15 seconds at 5.685 MHz and 15 seconds at 8.975 MHz. It made five circuits at 1000, 2000, 3000, 4000 and 5000 feet above ground level).

> Mercaptans such as ethyl mercaptan, thiophenol and octyl mercaptan are converted by aqueous alkaline hydrogen peroxide to the corresponding disulphide. Yields are high. Acidified peroxide may be added so as to avoid cleavage of the disulphide by excess alkalinity and so that precipitation of the product may be assisted. In the case of branched chain high molecular weight mercaptans, some resistance to oxidation by hydrogen peroxide occurs.
>
> (Aqueous alkaline hydrogen peroxide converts mercaptans such as ethyl mercaptan, thiophenol and octyl mercaptan to the corresponding disulphides. Yields are high. Acidified peroxide can be added to prevent cleavage of the disulphide by excess alkalinity and to assist precipitation of the product. Branched-chain high-molecular-weight mercaptans resist oxidation by hydrogen peroxide.)

The moiré effect is dependent upon the difference between the spacing of the lines and the pitch of the holes. The alignment of the brightness waves is not necessarily in parallel with the picture lines and any form may be assumed by the waves. The modulation depth of the waves is dependent upon the diameter of the luminance spot on the screen. No moiré effect will be experienced if the spot is large or if the lines are wide.

(The moiré effect depends on the difference between the spacing of the lines and the pitch of the holes. The brightness waves do not necessarily run parallel with the lines and may assume any form. The modulation depth of the waves depends on the diameter of the luminance spot on the screen. There will be no moiré effect if the spot is large or if the lines are wide.)

Note, too, that in improving the passive versions, it was not necessary to introduce personal constructions. **Active writing does not have to be personal**.

I suspect it is the confusion of writing actively with writing personally that leads many teachers to enforce the 'rule' that scientific and technical reports should be written in past-tense, passive-voice, impersonal constructions. Their intention is reasonable: they wish to develop in their students a detached, 'objective' outlook appropriate to scientific activity. But proper detachment does not require roundabout, passive thought and expression. Let us encourage objectivity by all means; but since passive-voice constructions may distort meaning and increase ponderousness of expression, let us encourage incisive, active writing as the norm.

2.12 VERBS: IMPERSONAL VS FIRST-PERSON CONSTRUCTIONS

Excessive use of passive constructions seems artificial and uncomfortable to readers. On top of this, ambiguity and confusion are often added because passive constructions often do not specify clearly who was the actor or agent in an event. I misunderstood all the following extracts on first reading:

> Smith and Robinson have determined the extent of thermal degradation by heating for standard times at 220°C and then calculating the weight loss during this time, together with the changes occurring in the infra-red spectrum of the polymers. The colour of the polymers has also been reported, as it gives some guide to the extent of the reaction. This stage of the investigation is nearing completion, and it is planned to examine the products of pyrolysis in the region of the exotherm using gas chromatography.

> Green and White considered that the relative post-mortem stretching of the intestine is independent of the size of the animal, and it is especially rapid if the intestine is freed from the peritoneum.
> After attempts had been made to measure the length of the pig intestine, it was realized that these measurements were of little value.

> The following investigation was begun as a result of recommendations by John Brown in the report 'Measurements of Base-plate Profiles' (No. AB 123). In that report he noted that... In other work recorded in the above report it was found that the top edge suffered no appreciable dimensional change, even after several years' service. However, new base-plates often were outside specified tolerances. It was decided that the measurements should be made on a free-standing atmospheric rig, using air as the working fluid. Six plates were to be tested under varying...

In each of those extracts, I missed a change of theme that was masked by use of impersonal constructions.

In discussion of the first extract, I discovered that *This stage of the investigation is nearing completion* did not refer to Smith and Robinson's work but to the author's own work. If he had written *In our investigation, we have almost completed...and we are planning* there would have been no confusion.

In the second example, it is not clear who attempted to measure the length of the pig intestine; it looks as if the second paragraph continues the account of Green and White's work. If the writer had begun his second paragraph with the words *After we had attempted...we realized*, we should have followed the transition from previous work to new work confidently and comfortably.

The third extract seemed at first to be entirely devoted to describing previous work by John Brown. In fact, the penultimate sentence begins to describe decisions taken by the report writer in designing his own work. The transition could have been completely clear if he had written *I decided that I would make my measurements*.

It would surely have been unreasonable to accuse these writers of presumptuous self-glorification for using personal constructions in these examples. There would have been no lack of humility, or distortion of scientific fact; we should simply have been grateful to have the contrast between the earlier work and the new work clearly shown. The gain in clarity would have been considerable.

Detachment vs vagueness

The need to preserve 'proper' scientific detachment and humility is often stressed by teachers who will not accept personal, active writing in scientific and technical papers. They contend that readers of scientific and technical papers are interested primarily in the scientific facts, not in who established them; the use of *I* or *we* is an unwarranted, immodest introduction of a specific agent into the account.

I readily agree with the need for clarity of perception and for balanced, detached comment. But I have argued already that there are points in scientific and technical papers at which specification of an agent is essential to the clarity of the account. At the beginning of this section, I gave three examples from introductions to papers (page 60). Here are more examples, this time from discussion sections:

Some study is being undertaken at the moment into the application of narrow aisle trucks in the warehouse.

Consideration has been given to...

Attempts have been made to prepare...

Recommendations are made that...

The suggestion is made that...

It is recognized that these will be subject to change.

It is the intention that these methods be subjected to further refinement.

It is believed that eight radiators will be sufficient.

It has been remarked previously that...

It has been found by experiment that...

It has been established previously that...

It is claimed that...

It is thought that...

It is felt that...

It has been observed that...

Whose ideas or work is being described here? Who is responsible for the opinions being expressed? Who is making or has made the claims or remarks? Who thinks or has thought, who feels or has felt or has made the observations? Readers do not know. *It...that* constructions, in particular, often create vagueness, or a sense of hesitancy or evasion. Impersonal writing of this type should be discouraged, not enforced.

I contend that, in the contexts in which those extracts appeared, the clarity of the messages would have been increased considerably if the author(s) had written:

We are currently studying the use of narrow aisle trucks in the warehouse.

We have considered...

We have tried to prepare...

We recommend that...

I suggest that...

I recognize that these will be subject to change.

We intend to refine these methods further.

I believe that eight radiators will be sufficient.

I have remarked previously that...

In previous experiments, we have found...

We have established previously that...

We claim/believe...

I think...

I feel...

We have observed/noticed...

My over-riding concern in all this advice is for clarity and accuracy of communication. Automatic adoption of *It...that* constructions frequently reduces both clarity and accuracy. Though such constructions are legitimate and natural features of English, I suggest you should use them sparingly and with precise intention. The main reason for this suggestion is that *It is believed that...* does not normally have the same meaning as *I/We believe*. In general usage in English, an *It...that* construction signifies either a generally held belief in the community, or a collective responsibility or statement:

In the USA, it is believed that addition of fluoride to the water supply...

It has been found that the diffusion-cell systems and experimental techniques first used in the 1982 studies by Brown and Smith are very reliable [1, 2, 3].

It is therefore at least confusing and at worst dishonest for writers to use *It...that* constructions when they mean *I/We think*. Note the confusion in the first version of the text that follows, and the greater clarity of the second version:

It has been found that the diffusion-cell systems and the experimental techniques first used in the 1982 studies by A and B are very reliable [3, 4, 5]. However, it was found in recent experiments that it was impossible to obtain results equivalent to..., using cell-system X...

It has been found that the diffusion-cell systems and the experimental techniques first used in the 1982 studies by A and B are very reliable [3, 4, 5]. However, in recent experiments, using cell-system X, we were unable to obtain results equivalent to...

Regrettably, writers get locked into an *It...that* frame of mind, and begin to use it when it is totally unnecessary:

It has been observed by Green and White that...
(Green and White have observed...)

it was considered by Muguruma (1961) that...
(Muguruma (1961) considered that...)

In an earlier paper (Smith 1961) it was shown that...
(In an earlier paper (Smith 1961) I showed...)
(In my 1961 paper, I showed...)

Equally regrettably, *It...that* constructions are sometimes used deliberately to suggest wider support or acceptance for a statement than is warranted. Dr I.J. Good has alerted scientific readers to the real meaning of many such impersonal constructions [10]:

It is clear that much additional work will be required before a complete understanding...
(= I don't understand it.)

It has long been known that...
(= I haven't bothered to look up the original reference...)

I have no quarrel with genuine attempts to present facts with detachment and diplomacy; but I have no time for the spurious 'objectivity' that produces silly circumlocutions such as:

It was speculated that...
(I thought/speculated/guessed...)

It is highlighted that...
(We would stress that...)

It is strongly considered that...
(I firmly believe/consider...)

It was felt subjectively that smooth operation was not likely.
(I thought smooth operation was unlikely.)

It is suggested by the author that...
(I suggest that...)

The results lead the present authors to believe that...
(The results lead us to believe that...)

Before the experiment, the author was very uncertain how to...
(Before the experiment, I was very uncertain how to...)

It is the authors' preference to install...
(We prefer to install...)

The author was glad to avail himself of the opportunity to make vibration measurement...
(I was pleased to take the opportunity to...)

It is the writer's opinion that...
(In my opinion...)

No details are available to the present writer on the specification of...
(I have no details of the specification of...)

In the present author's experience...
(In my experience...)

The author's wife, Mrs John Smith, typed the thesis...
(My wife typed the thesis...)

Once again, I am pointing out that the 'rule' is bad not because it advocates the use of a particular form but because it advocates the use of that form almost to the exclusion of all others. Impersonal constructions have a proper and useful place in our natural use of English. But they are not automatically interchangeable with personal constructions.

Consider the following examples: the personal and impersonal statements do not have identical meaning:

I do not accept the idea...
One cannot accept the idea...
The idea cannot be accepted...

I suspect that...
It is possible that...
A suspicion may arise that...

We do not understand how the chlorine reacts with...
It cannot be understood how the chlorine reacts with...

...gave one of the strongest wiremarks we have ever seen.
...gave one of the strongest wiremarks ever experienced.

...and we have information that...
...and there is information that...

The last three pairs were changes from personal to impersonal constructions made by senior staff 'editing' young scientists' texts. The editing was harmful. Writers must not be required artificially to force all their thoughts into an impersonal mould: there is real danger that they will distort their intended meaning.

Most scientists, engineers, or computer specialists would be horrified at the thought of unrecognized inaccuracy in their experimental work; yet an astonishing number cheerfully ignore vaguenesses and approximations of meaning that creep in as they strait-jacket their ideas into stereotyped patterns of expression.

'General-purpose' verbs

The habitual use of impersonal constructions has an insidious inflating effect. Writers who will not (or feel they **must** not) write:

We sampled the ions from the plasma by...

I removed the coating with alcohol.

We did not inspect the burners regularly.

could write in simple passive form:

The ions from the plasma were sampled by...

The coating was removed with alcohol.

The burners were not inspected regularly.

But time and again they yield to temptation to take a further step and expand these statements to:

Ion sampling from the plasma was achieved by...

Removal of the coating was effected by the application of alcohol.

Regular inspections of the burners were not carried out.

In taking this extra step, they not only change the construction from personal to impersonal and from active to passive: they also introduce colourless 'general-purpose' verbs carrying abstract nouns. They no longer *sample, remove* and *inspect*; they *achieve, effect* and *carry out*. And the focus of their statements is no longer on *ions*, the *coating* or the *burners*; it moves to the vaguer abstractions *sampling, removal* and *inspections*.

Undoubtedly, there are many occasions when we want to focus on *sampling* rather than on *ions*, or on *inspections* rather than on *burners*. In such circumstances, as I have argued earlier (pages 54 to 59), the use of a passive construction is a proper and valuable way of moving the emphasis of a sentence. I repeat that I am not condemning **all** use of passive constructions: I am simply anxious to show the pernicious inflating effect of using impersonal passive constructions with 'general-purpose' verbs as the rule rather than as the exception.

Much of the ponderous monotony of technical writing is contributed by this practice. Just consider how much time technical people apparently spend *performing*:

Measurements of reflection and transmission are performed.
(Reflection and transmission are measured.)

The integration of X with Y may be performed.
(X may be integrated with Y.)

An experiment was performed on a small scale to ascertain the influence of X, which resulted in...
(In a small-scale experiment to ascertain the influence of X,...)

Our facilities were inadequate to perform the experiment.
(Our facilities were inadequate for the experiment.)

Our present measurement of surface area is performed by...
(At present, we measure surface area by...)

Calculations of the yield were performed which revealed...
(We calculated the yield and found...)

A second trial was performed...
(In a second trial, we...)

and *conducting*:

Tests have been conducted on the pavements.
(The pavements have been tested.)
(We have tested the pavements.)

A test was conducted to discover the value of X.
(In a test, the value of X was...)

Another study of the roof was conducted twelve months later.
(The roof was studied again twelve months later.)
(We studied the roof again twelve months later)

Trials were not initially conducted on operational aircraft.
(At first, there were no trials on operational aircraft.)

and having *experiences*:

Unusual shrinkage of the diaphragm plaque was experienced.
(The diaphragm plaque shrank unusually.)

A high number of breakdowns was experienced in March.
(There was a high number of breakdowns in March.)

Quantities in excess of one litre were experienced.
(More than one litre was produced)

Greater success was experienced on the plant in reducing the free acid concentration than was experienced in the laboratory experiments
(On the plant, the free acid concentration was reduced more successfully than in the laboratory experiments.)
(On the plant, we reduced the free acid concentration more successfully than in the laboratory experiments.)

They are great *carriers* too:

Volume control of the individual speakers is carried out by a switch.
(A switch controls the volume of the individual speakers.)

Regular inspections of the burners are not carried out.
(The burners are not inspected regularly.)

Periodic cleaning and lubricating of the fan filters should be carried out every six months.
(The fan filters should be cleaned and lubricated every six months.)

To date, most work has been carried out on sheets of...
(To date, most work has been on sheets of...)

Daily measurements of levels are carried out.
(Levels are measured daily.)

Installation of the units will be carried out by works personnel
(Works personnel will install the units.)

A final test was carried out consisting of running the rig contin-
uously.
(In a final test, the rig was run continuously.)
(In a final test, we ran the rig continuously.)

Refinement of this plan was carried out.
(This plan was refined.)
(We refined this plan.)

Load-cycle tests were then carried out.
(Load cycles were then tested.)
(We then tested load cycles.)

Measurements of electron and ion densities in argon afterglows were
carried out.
(Electron and ion densities in argon afterglows were measured.)
(We measured electron and ion densities in argon afterglows.)

and they *accomplish* great things:

Vacuum sealing of the waveguide itself was accomplished using
quartz windows:
(The waveguide itself was vacuum-sealed using...)

Heat-treatment of the film was then accomplished.
(The film was then heat-treated.)
(We then heat-treated the film.)

Expressions such as *was performed, were conducted, were experienced, were carried out, was achieved, was shown, were effected, were
observed, were accomplished, resulted,* and *occurred* are desperately over-
worked in technical writing because technical writers are reluctant to
write directly and personally. Notice that it was possible to remove
most of the roundabout expressions without introducing a personal
construction; but would introduction of personal constructions necessarily have been undesirable? Is it true that occasional introduction of
personal constructions *always* strikes technical readers as obtrusive or
immodest?

Acceptability of first-person style
The reactions of the respondents to my surveys deny this. The majorities **preferred** texts that used a judicious mixture of personal and

impersonal constructions (pages 167 to 197). There was some confused thinking among the minority:

> I would have given version H (using personal pronouns) first prefer-ence had the 'we's' been eliminated, e.g. by 'it was felt', not 'we thought'...

> Version H (using personal pronouns) would have been easily first if written in the third person...

Many who were reluctant to vote in favour of texts that included personal pronouns acknowledged that those texts were very clear. Many admitted, too, that their reaction against those texts was prob-ably due to brainwashing during their education, not to any rational objection (pages 195 to 197).

Let me stress that in advocating the use in writing of the personal, active phrasing of serious discussion, I am not advocating the use of the casual, inexplicit shorthand of person-to-person chatter. I am not advocating that the account should be over-personalized, sensational-ized or trivialized. Over-indulgence in personal constructions is just as distorting as non-indulgence: to sound self-congratulatory is as bad as to pretend you were not involved at all. And I am not suggesting that the technical content should be distorted in any way. I **am** advocating that writers should write naturally and economically, without affect-ation of a special 'scientific' style. They should come as close as possible to the natural mixture of constructions and the natural balance and rhythms of comfortable everyday speech.

My advice has sometimes been misrepresented as encouragement to writers to claim credit for work they did not do themselves. That might be true if I advocated that all impersonal, passive construction should be replaced by *I* or *we*. I make no such suggestion. If much work has been done on a project by someone other than the writer or the writer's group, then it is entirely appropriate to write *much work has been done* or *in many experimental programmes it has been found that*, with or without attribution to a particular agent. I advocate simply that writers should make statements active wherever possible. For example *installation of a noise barrier has been carried out by Smith and Jones* should be converted to *Smith and Jones have installed a noise barrier*. In particular, where writers have been personally involved in some work, they should not use cumbersome passive constructions simply to avoid saying *I* or *we*. They should make clear distinctions between their own observations, beliefs and deductions, and those of other writers.

Once again, I am suggesting that it is the blanket operation of the 'rule' that is harmful. Writers should be taught that the patterns of expression they will naturally require will probably vary in different parts of their reports and papers. Introductory sections will be the most varied in tense, voice and person. Descriptions of procedures will be mainly past tense and impersonal, but should be kept as active as possible. Conclusions and recommendations will vary in tense and voice. Since they convey the writer's conclusions and recommendations, they can quite properly be personal; but both will often be most effectively expressed as lists:

Conclusions
1. A policy of part-life-rework for ABC engines is essential.
2. The engine bay at Someplace works could cope with the work involved in operating such a policy.
3. A part-life-rework policy would necessitate the issue of revised modification standards and service operating procedures.

Recommendations
1. that the guard-leads be screened in plastic (para 5.3);
2. that the gear ratio to the shaver be reduced to 10 : 1 (para 6.4);
3. that the measuring head be fitted with a dust shield (para 7.2).

I have been told that it is essential to train students – especially research students – to write impersonally and passively because most editors of professional journals will not accept articles that contain personal constructions. That simply is not true. In a walk round the Periodicals Library at the University of Wales in Cardiff, I took the current copies of 81 scientific journals at random from the shelves. I found articles using *I*, *we* or *our* in 74 of them:

Acta Crystallographica
Advances in Physics
American Journal of Physics
Biochemical Journal
Biological Conservation
British Journal of Industrial Medicine
British Journal of Ophthalmology
British Journal of Nutrition
British Journal of Pharmacology
British Medical Bulletin
British Medical Journal
Bulletin of Entomological Research
Chartered Municipal Engineer
Chemical Engineer
Chemical Engineering Science
Computer-Aided Design
Computer Journal
Contemporary Physics

Cryogenics
Econometrica
Electrical Review
Electronics and Power
European Polymer Journal
Experimental Eye Research
Faraday Discussions of the Chemical Society
Faraday Symposia of the Chemical Society
Hydrographic Journal
IEEE Transactions
Inorganic Chemistry
International Journal of Electronics
International Journal of Engineering Science
International Journal of Mechanical Sciences
International Journal for Numerical Methods in Engineering
Journal of the American Chemical Society
Journal of Applied Ecology
Journal of the Chemical Society, Faraday Transactions
Journal of the Chemical Society, Perkins Transactions
Journal of Ecology
Journal of Environmental Management
Journal of Fluid Mechanics
Journal of the Geological Society
Journal of the Institute of Mathematics and its Applications
Journal of Materials Science
Journal of the Mechanics and Physics of Solids
Journal of Navigation
Journal of Pharmacy and Pharmacology
Journal of Physics
Journal of Physiology
Journal of the Royal Statistical Society
Journal of the Science of Food and Agriculture
Lancet
Manufacturing Technology
Microelectronics Journal
Mikrochimica Acta
Nature
Ocean Management
Operational Research Quarterly
Philosophical Magazine
Physica Status Solidi
Polymer
Polymer Engineering and Science
Proceedings of the American Society of Civil Engineers

Proceedings of the London Mathematical Society
Proceedings of the Royal Society of London
Progress in Materials Science
Psychopharmacology
Quarterly Journal of Mechanics and Applied Mathematics
Quarterly Journal of the Royal Meteorological Society
R & D Management
Radio and Electronic Engineer
Spectrochimica Acta
Transactions of the American Society of Mechanical Engineers
Transactions of the Institution of Chemical Engineers
Water Resources Research

The remaining seven seemed entirely impersonal:

Journal of the Institution of Water Scientists
Journal of Mechanical Engineering Science
Machine Tool Design and Research
Powder Technology
Proceedings of the Institution of Civil Engineers
Strain
Water Pollution Control

Two notes must be added:

1. The omission of a particular journal from these lists is not a slight on that journal: it implies simply that I did not get round to it in my random selection. I gave up browsing after two hours, feeling that my point was amply supported.
2. I do not know whether the seven journals that seemed impersonal have a deliberate policy to be so. They, too, may be prepared to accept good personal writing if they receive it.

The way in which we permit ourselves to select language is of outstanding significance, because it goes a long way towards establishing our over-all 'mental set' – our customary habit of attack on our daily tasks of thought and expression. Are we to think as directly, positively and flexibly as possible, or are we to approach every statement in a stereotyped, roundabout, non-committal, self-protective way? Certainly, there are times when we have to write diplomatically. Certainly, there are many contexts and topics – for example, descriptions of machines or reactions – that do not naturally call up personal constructions. But in most scientific papers and reports, we are concerned to be as direct, explicit and economical as we can. In those circumstances, we should write in a natural, comfortable mixture of personal and impersonal constructions, using active verbs as our main mode of expression, and interweaving passive verbs skilfully to change the balance and emphasis of what we want to say.

2.13 VERBS: IMPERSONAL VS SECOND-PERSON CONSTRUCTIONS

The discussion of first-person style in section 2.12 relates mainly to the writing of reports or papers; it is not so relevant to the writing of guides or manuals. A report or paper is essentially a personal statement made by one or more writers, albeit sometimes on behalf of a company or consultancy. In such a text, an entirely proper orientation for the writer(s) is to use *I* or *we*, giving a straightforward account of what he/she/they/the consultancy have found, think, or suggest. A guide or manual is an attempt by a manufacturer or supplier to offer a buyer or user information that will be helpful in the operation of a product, process, or service. It is rarely (if ever) appropriate for the writer or writers of a manual to speak personally to the readers. It **is** reasonable occasionally for the writers to use a first-person-plural style, when presenting suggestions on behalf of the company:

> If you find that maintaining so many files is difficult to manage, we suggest that you...

but in descriptive and explanatory texts, even this is rarely appropriate.

Tactics for writing descriptions, explanations, and instructions are discussed in detail in Chapters 4 and 5. In this general chapter on style, I want just to discuss the general benefits of using a second-person style (using *you, your*) in guides and manuals, instead of impersonal, passive style.

If you use second-person constructions, you are able to write more directly, often in an involving, 'user-friendly' tone. Impersonal style seems more distant, and it tempts you into:

- passive constructions;
- awkward use of *the operator* or *the user*, which in turn often leads to sexist wording;
- grammatical errors (faulty use of constructions based on participles and infinitives).

Passive constructions

The following text is not difficult to understand; but it has some avoidable passive constructions, and the word *user* is obtrusively repeated:

> XY-MAIL is XY's electronic mail system for information distribution. Messages can be sent to any user located at any terminal in an

XY-NET network. When a message arrives in a user's in-tray, the user will be notified, and can choose whether to forward it to another user, print it, just read it, or store it for later retrieval.

A second-person version of the same information removes the passive constructions, and helpfully distinguishes between *you* the operator of the system, and other *users*:

XY-MAIL is XY's electronic mail system for distributing information. You can send messages to any user at a terminal in a XY-NET network. When a message arrives in your in-tray, you are notified, and you can choose whether to forward it to another user, print it, just read it, or store it for retrieval later.

Sexism

If you attempt to write impersonally, beginning with *the operator* or *the user*, you will often run in to problems in selecting pronouns. If you begin with *the operator* or *the user*, it is almost inevitable that you are tempted eventually into 'sexist' writing (use of the masculine terms he/his/him when the operator/user could equally well be male or female):

By means of the flow-chart, the operator is guided along a strictly logical route, prescribed by the manufacturer. Even an operator without specialist knowledge of the equipment can find his way around it, and locate faults...

Conversion of the text to a second-person style is not the only tactic open to you to avoid this unacceptable sexism. You can turn the whole text into impersonal, passive constructions; you can use the optional pronoun *he/she*; or you can convert the discussion into plural terms, and use *they* when you need a pronoun. All of these tactics have disadvantages. Repeated use of passives becomes roundabout and dull. Repeated use of *he/she* becomes obtrusively clumsy. Conversion to plural terms is sometimes inappropiate in a discussion of an activity involving only one person.

Use of second-person style is often the best tactic, especially for descriptive and explanatory texts in which you discuss several possible conditions, outcomes, or courses of action:

The flow-chart guides you along a strictly logical route that we prescribe. Even if you have no specialist knowledge of the equipment, you can find your way around it, and locate faults...

If you have a hard disk of 32 Megabytes or less, you have the option of dividing it into more than one volume. (Your computer would see each volume as smaller, separate disk drives on which different information of operating systems could be placed.) If, however, you have a hard disk of more than 32 Megabytes, you **must** divide it into multiple volumes...

Some writers worry that the use of *you* will sometimes be inappropriate because the reader is not always the person who will be the operator of an item of equipment or a process. They contend that there are sometimes three parties involved in the discussion: the writer, the reader, and a third party – the operator.

Certainly, there are occasions when the person reading the manual is not, and will not be, the operator of the machine or process. For example, the manager or system administrator in a computing department may be reading an Operator's Guide that will be used only by his or her staff. However, in those circumstances, the reader usually reads the text as though he or she was assuming the role of the operator. Indeed, a good manager who is evaluating the suitability of a text for use by his or her staff should try to read that text with the eyes and expertise of those staff members. It is therefore entirely reasonable to write that Operator's Guide as though you were addressing the operator (or potential operator) directly.

As so often in this book, my advice reflects the need for rigorous thinking about the aim, audience and context for a text. In a System Administrator's Guide, it may be necessary to distinguish carefully what **you** (the administrator) have to do, from what **the operator** has to do. In an Operator's Guide, it may be necessary to exchange the terms, and distinguish carefully what **you** (the operator) have to do, from what **your System Administrator** has to do.

Grammatical errors

Perhaps the greatest benefit of adopting a second-person style – talking directly to your reader – is that it helps you avoid grammatical errors that frequently slip in when writers are writing impersonally.

The most common of these errors is the misrelated participial construction. In English, we have a grammatical rule that a participle (formed usually by the addition of -*ing*, -*ed*, or -*d* to the infinitive form of a verb) relates to the noun or pronoun that precedes it:

The operator, having loaded the program, types in the values...

Four cylinders, packed with Product X, are linked in series...

If there is no noun or pronoun at the beginning of the sentence, the participial group is interpreted as relating to the subject of the main statement that follows:

After discharging the load from the hopper, the operator removes the pan...

Although filled with water, the tubes are still sufficiently...

Writers working in an impersonal style frequently lose sight of this rule, and produce nonsensical statements:

After discharging the load from the hopper, the pan is removed by the operator...

Although filled with water, the technician still has sufficient...

After depressing a button in the centre of the operating handle, the door can be opened from the inside.

Before installing the battery, the function switch must be in the OFF position.

After loading the test cassette, the loading menu gives a choice of entering one of four systems.

When using these fixed targets, accurate calculations must be made...

If you are thinking in a second-person style, your pattern of thought helps you to avoid such errors:

After depressing a button in the centre of the operating handle, you can open the door from the inside.
or:
After you depress..., you can open...

Before you install the battery, the function switch must be in the OFF POSITION.
or:
Before you install..., you must place the function switch...

After you have loaded the test cassette, the loading menu gives you a choice of entering one of four systems.
or:
After you have loaded..., you are given a choice...

When you are using these fixed targets, accurate calculations must be made...
or (preferably): When you are using these fixed targets, you must make accurate calculations.

Second-person style helps you avoid misrelating participial constructions at other positions in sentences, too:

Not The wailing tone stops after pressing the cut-out switch.

But The wailing tone stops when you press the cut-out switch.

It also helps you avoid mis-relating infinitive constructions:

Not To monitor the conversation, the loudspeaker must be operated.

But To monitor the conversation, you must operate the loudspeaker.

Not To configure a channel, the register has only to be loaded once.

But To configure a channel, you have to load the register only once.

However, let me temper my advice on using second-person style by adding two warnings:

- **over**-use of *you* can become as tiresomely obtrusive as over-use of any other element of language, so it must be used judiciously;
- use of *you* can tempt you into an abbreviated style, using *you'll*, *you're*, *you've*, which many readers find inappropriately casual.

In section 2.15, I shall discuss the importance of **tone** in technical writing. I shall stress that readers have a sense of what is 'about right' in choice and use of language for technical writing. I want simply to point out here that some readers react against the use of *you* at all (witness the results of my survey of style for manuals, Appendix F, which showed a minority of readers preferred other styles). Also, many readers feel that the use of abbreviated expressions such as *you'll* and *you're* (and *can't* and *won't*) are inappropriately 'sloppy' in a serious technical text.

I do not know of any hard evidence of the acceptability or unacceptability of an abbreviated style; I can say only that in my experience up to the time of writing this book (1991), a majority of readers in technical contexts react unfavourably towards such a style. I do not believe the reaction is violent, but it is a factor that you should keep in mind as you make your choices of language for technical writing.

2.14 PUNCTUATION

I have discussed punctuation with many hundreds of scientists, engineers, computer specialists, and professional technical writers. Time and again, I have been told that no-one has explained to them the rationale and system behind punctuation in English. That is a severe condemnation of the English-teaching profession. In partial redress of the situation, here is a brief rationale for careful punctuation in technical writing. A full account of the system would almost fill the whole of this book, so I have mentioned in this section just one or two points that are vital to clear writing and easy reading. For detailed guidance on punctuation for technical writing, see my book *Full Marks* [11]. For discussions of punctuation in general contexts, I recommend the short, sections in *Modern English Usage* [12] and *The Complete Plain Words* [13], or the short books *Mind The Stop* [14], by G.V. Carey and *You Have a Point There* [15] by Eric Partridge.

Omission of punctuation is distracting and confusing. A constant irritation that I find in technical writing is the omission of hyphens. Sometimes the collocation of words without hyphens makes such nonsense that readers have no difficulty in detecting the intended meaning:

> ...made counts of the bacterial loads in the air. He observed that bacteria carrying dust particles decreased in concentration as the humidity decreased.
> (bacteria-carrying)

> ...can be present in boron containing steels...
> (boron-containing)

But readers should not have to work out the correct relations between words because writers have failed to signal them clearly. It should not be necessary to read the following examples twice in order to be sure what they mean:

> Additive free
> (Additive-free)

> Before one can discuss plans to cut down noise one must isolate the noise creating factors.
> (noise-creating)

Taking off gases from the esterifiers...
(off-gases)

To maintain the system's low current measurement capability...
(To maintain the system's low-current-measurement capability...)
(To maintain the system's ability to measure low currents...)

Absence of commas, especially to mark off words and clauses at the beginning of statements, is another regular source of ambiguity and irritation. In each of the following pairs, the writer wrote the first but meant the second:

Frequently adjusted totals need to be scrutinized.
(Frequently, adjusted)

Should any burner fail to ignite its respective section will revert to 'purge' and in this way the...
(fail to ignite, its respective)

As the truck passed the vibrations were noticeable to anyone standing on the bridge.
(the truck passed, its vibrations)

As the machine develops the forms we use to record data from past projects will be amended.
(the machine develops, the forms)

In the event, therefore, of failure of any of these components to operate the respective motor will not start until the fault has been rectified.
(components to operate, the respective motor)

In all these examples, the writers have put their words in suitable sequences; but they have failed to signal to us explicitly two important categories of information:

- the relationships they intend between words and word-groups;
- the boundaries between significant word-groups.

Signalling relationships

As an example of punctuation marks being vital to clear signalling of relations between word-groups, consider the statement:

He observed that bacteria carrying dust particles decreased in concentration.

The words as they stand indicate the following relationships in the word-group beginning with *that*:

> *bacteria* is the subject
> *carrying dust particles* is an adjectival group describing the bacteria
> *decreased* is the verb, telling us what happened to the bacteria

When a hyphen is placed between *bacteria* and *carrying* the relationships between the words are changed:

> *particles* is the subject
> *dust* is an adjective describing the particles
> *bacteria-carrying* is an adjectival group describing the particles
> *decreased* is the verb, telling us what happened to the dust particles

In discussion with the writer, I discovered that the latter relationship was the one he intended.

Signalling boundaries

As an example of punctuation marks being vital to clear signalling of boundaries between significant word-groups, consider the sentence:

> As the machine develops the forms we need to record data from past projects will be amended.

When I read those words for the first time, I took the group *the forms* to be the object of the verb *develops*. As I searched for the first significant group in the string of words, my eyes and mind 'closed' round *As the machine develops the forms*. They went on to add *we need* and *to record data from past projects*, so that I assembled in my mind the interpretation:

> As the machine develops the forms we need to record data from past projects...

But then I arrived at *will be amended*. Immediately, it was clear that I had misinterpreted the first part of the sentence. I had missed the intended boundary between *develops* and *forms*:

> As the machine develops, the forms we need to record data from past projects will be amended.

In discussion with the writer, I discovered that this was indeed the meaning he intended to signal.

Of course, my mind made the necessary adjustments in much less time than it has taken me to discuss those two examples of unpunctu-ated text. But on each occasion, the inexpert writing disturbed for a moment my concentration on the argument the writer was presenting.

Is punctuation really important?

You may be tempted to ask (as I *have* been asked) whether such a small moment of disturbance is really very important. My answer is that, **by itself**, a single omitted comma is not a disaster; but a text that frequent-ly omits punctuation becomes a struggle to read.

Supporters of 'light' punctuation (which usually seems to mean the use of full stops, but not much else) sometimes argue that the careful wording of sentences is enough to make clear the meaning and tone intended. Of course, the position occupied by each word in a sentence is an important indication of the role to be played by that word. But as my examples show, accurate positioning of words is not enough by itself: to decode messages easily and accurately, readers need addit-ional information from punctuation marks.

Punctuation marks are integral parts of the code on which written communication is based. That code consists of three sets of agreements or rules:

- about **individual words** (lexical rules);
- about **word order** (grammatical rules);
- about how to signal **grammatical/logical information** and **rhetorical information** (in speech, rules of intonation and stress; in writing, rules of punctuation).

We have agreements about which sounds or symbols we shall use to express things, actions, and ideas (Examples: *word, write, communica-tion*). We have agreements about the order in which we shall arrange words to make statements (Example: *In order to communicate, I speak or write words in an agreed sequence*). But those alone are not enough. We cannot just string words together in a continuous outpouring of sound or as an unmarked string of hieroglyphics on paper. We have to show the relationships between the words. The following sets of sentences have the same words, but do not have the same meanings:

Set 1
1. Hit the man using your brother's chair.
2. Hit the man, using your brother's chair.
3. Hit the man using your brothers' chair.

Set 2

1. Insert the old disk into the disk drive with the notch at the bottom and the label on the left.
2. Insert the old disk into the disk drive, with the notch at the bottom and the label on the left.
3. Insert the 'old' disk into the disk drive, with the notch at the bottom and the label on the left.

In speech, we emphasize differences of meaning by changes of tone, changes in inter-word spacing, and by placing stress on different syllables and/or in different positions in a sentence. In writing, we have to supply marks on paper in place of the voice signals we would use if we were speaking. As those sets of examples show, the presence **or the absence** of a mark can be a signal that indicates the part to be played by a word-group in a sentence.

In Set 1, the function of the word-group *using your brother's chair* is changed by the presence or absence of the comma. In sentence 1, it defines which man we should hit (its function is adjectival, modifying *man*); in sentence 2, it tells us how we should hit him (its function is adverbial, modifying *hit*). In sentence 2, the placing of the apostrophe acknowledges that we have just one brother; in sentence 3, it acknowledges that we have more than one brother.

In Set 2, the function of the *with...* group is changed by the presence or absence of the comma. In sentence 1, it tells us that the disk drive has the notch at the bottom and the label on the left (its function is adjectival, modifying *drive*); in sentence 2, it tells us how to insert the disk (its function is adverbial, modifying *insert*). In sentence 3, we see the use of inverted commas (quotation marks) to fulfil a rhetorical function – to express a certain tone in addition to the basic meaning of *old*. The presence of the inverted commas signals that the writer is using the word *old* in an unusual way: he or she wishes to signal that the disk is not truly old, but that *old* is the term allocated to it for the current discussion.

These examples emphasize that punctuation marks are not optional extras. If anyone challenges the view that punctuation marks are an integral part of our code, invite him or her to read the following text aloud, unhesitatingly, and with immediate, clear expression:

What are directories?

Like files directories are containers but instead of text or other data directories contain files in addition directories are hierarchically organized that is a directory has a parent directory above and may also have subordinate child directories below each child directory in turn may contain other files and may also have child directories and so on because they are hierarchically organized directories provide a logical way to organize files.

For comparison, here is the text as the writer wrote it, with meaning signalled clearly:

What are directories?

Like files, directories are containers. But instead of text or other data, directories contain files. In addition, directories are hierarchically organized; that is, a directory has a parent directory 'above' and may also have subordinate child directories 'below'. Each child directory, in turn, may contain other files, and may also have child directories; and so on. Because they are hierarchically organized, directories provide a logical way to organize files.

The full impact of punctuation and paragraphing is difficult to demonstrate in short passages, even in passages such as were used in the surveys I mentioned in Chapter 1. Also, as I constructed those surveys, I decided that it would be unfair to load the scales heavily against some versions by deliberate mis-punctuation, which would have made comprehension difficult. My aim was to stop any of the versions seeming obviously 'wrong', confused or unacceptable because of glaring faults of organization, grammar, punctuation or spelling.

Nevertheless, the survey passages are long enough to show how skilful punctuation and paragraphing contribute to the clarity and readability of a text. Readers who are seeking guidance on the systematic use of punctuation marks may find it useful to begin by studying those passages.

The best punctuation

In this section, I have been commending full use of punctuation marks: but I am not encouraging over-indulgence. G.V. Carey's philosophy, expressed in *Mind the Stop* [16], is a good one to have in mind:

Stops should be used as sparingly as sense will permit: but in so far as they are needed for an immediate grasp of the sense or for the avoidance of any possible ambiguity, or occasionally to relieve a very lengthy passage, they should be used as freely as need be. The best punctuation is that of which the reader is least conscious; for when punctuation, or the lack of it, obtrudes itself, it is usually because it offends.

2.15 TONE: IN HARD COPY AND IN ON-SCREEN TEXT

In most of this chapter, I have concentrated on accuracy and readability – on how to make meaning as clear as possible, and how to make text as easy as possible to 'digest'. However, before I finish this discussion of 'style in general', I want to mention the need to pay attention not only to clarity and readability, but also to the tone of what we write.

Tone is created both by what we say and by the way we say it. I expect you would have thought it tactlessly patronizing if I had begun this book or this section by saying, 'I congratulate you on buying and reading this book!'. Yet I met just that patronizing attitude and tone when I read the first page of the book entitled *Getting Started* that was provided for me when I bought my personal computer:

> We congratulate you on the purchase of the ABC personal computing system.

I have a *Beginner's Guide* to a software package, which begins:

> Welcome to the world of ABC...

And I have a *Configuration Guide* that tells me on its first page:

> This manual, together with its companion volume XXXX, represent a large step forward in the quality of system generation documentation.

My initial reaction on reading all of those openings was one of distaste. I have asked many other readers for their reactions, and they, too, reacted adversely to the tone of all three.

Why? The adverse reaction to all three openings seems to stem from a feeling that there is a self-congratulatory 'flavour' to the writing. When we approach 'sales' literature, we expect to have to tolerate a little arrogance, a little self-publicizing, from the seller of the goods we are considering buying. We have become accustomed (though perhaps not resigned) to the drum-beating, chest-inflating tone of advertising literature. But when we approach documentation that supports technical products, we are not looking for sugary salesmanship or self-satisfied 'confidence': we are anxious to acquire technical information quickly and comfortably. We expect to be addressed in an adult, serious (though not unduly solemn) manner.

We are therefore surprised by the tone of this beginning to an *Installation Manual*:

Your new computer has just arrived. You're looking for some easy instructions to get it set up and running, and you've turned to this...(manual)...for help. You've come to the right place.

Unfortunately for the writer, our surprise is not pleasant surprise. We feel that we are being addressed inappropriately. We feel we are being patronized.

I am sure that the writers of all these openings were well-intentioned: they were trying to write in a comfortable, 'user-friendly' way. They were trying to create a warm, conversational atmosphere, gaining the friendship and confidence of their readers.

For me, and for the majority of others to whom I have shown these extracts, the writers' efforts are unsuccessful. It is not easy to say precisely why, because we are in the realm of self-analysis, using introspection to try to examine our responses. Introspection is notoriously unreliable; but in the last few years, I have listened to enough responses from readers in the UK, the USA, and other European countries to be able to suggest some reasons why we should be very wary (at least!) of writing in the way demonstrated in the four extracts.

There is little doubt that audiences have a sense of the appropriateness of the tactics used in any attempt at communication. They have a sense of what is 'about right' on four measures:

- for an audience such as themselves;
- for a subject such as is being discussed;
- for a purpose such as theirs;
- for a context such as they are in.

Their judgement takes into account what is said, the way it is arranged, and how it is expressed. Writers can produce an unfavourable response from the audience by presenting too much information, or by presenting too little; by assuming too little expertise, or too much; by giving the audience too much guidance through the information, or too little; by adopting a brow-beating approach, or by apparently leaving readers to find their own salvation; by writing in a style that strikes readers as too formal, too ponderous, or unnecessarily involved; or by writing in a way that seems inappropriately flippant, inaccurately casual, or presumptuously 'forward'.

Choosing the appropriate material, arranging it conveniently, and expressing it suitably are difficult tasks. In this book, my main focus is on expression. But in discussing tone, it is impossible to ignore the inter-relation between what we say and how we say it. If we choose to make 'public-relations statements' to begin our technical manuals, it is difficult to prevent those statements emerging in public-relations language, and we run the risk of our readers finding the tone we create distasteful. If we choose to give massive displays of scholarly knowledge in challenge-proof academic papers or reports, it is difficult to prevent our language becoming complex and ponderous, and we run the risk of our readers resenting the efforts they have to make to disentangle the lines of our arguments.

'User-friendly' writing

Most of this chapter, 'On style in general', has discussed ways of preventing language used in technical writing becoming too complex, too heavy, too far detached from the natural vocabulary and rhythms of serious discourse in non-technical life. Inappropriate heaviness of style is one aspect of tone. In this final section, I want to warn against swinging too far in the other direction – towards an inappropriate, sometimes 'sickly' mode of expression, especially in technical manuals and 'on-screen' text.

I said earlier that I suspect the writers of the extracts I quoted were trying to be 'user-friendly'; they were trying to present their information in a way that readers – 'users' – would find easy to comprehend and agreeable to read. I believe the mistake in their tactics was to think that 'friendliness' lies in adding comments, quips, banter, 'light relief' to the text.

For me, and for virtually everyone with whom I have discussed this topic, a 'friendly' information source is one that presents the information I need in a format that I find attractive and easy to handle, in the amount and in a sequence that match my interest and convenience, in a style that I find clear, concise, and comfortable to read. Over-all, a text is friendly if it matches comfortably my intellectual and linguistic capacities, and my immediate interests/needs. When I am intent on acquiring technical knowledge or manipulative skill, I do not expect or like to be 'jollied along'. I swiftly get the sense that I am being treated like a child: the writer is confusing lack of expertise with lack of intelligence and/or lack of maturity.

Writers of 'on-screen' text seem particularly prone to this confusion, especially when they are writing for readers who are new to computing, or just new to a machine or package. The human:computer interface is different from the human:book interface. A computer seems to have a life of its own; a book is inert. A computer seems to respond actively; a book is passive. A computer seems to talk to its user; a book is silent. It is common to talk about 'man:machine dialogue'; I have never heard anyone talk about a 'man:book dialogue'. I must leave to psychologists detailed discussion of the psychology of the human: machine interaction. (If you are interested in that topic, I recommend, texts such as *Directions in Human-Computer Interaction* [17] and *Designing the User Interface* [18] by Ben Shneiderman, *Designing and Writing Online Documentation* [19] by William Horton, or *Text, ConText and HyperText* [20] edited by Edward Barrett.) I simply want to point out that many writers fail to recognize readers' sense of a much closer interaction with their computers. In a human:computer dialogue, the tone of the writer's voice comes through from the screen even more strongly than it does from a printed page.

Here is an example of an on-screen text. It adopts a tone that is intended, I am sure, to be user-friendly. It succeeded in irritating me extremely.

When I bought my personal computer, I received a self-teaching, 'hands-on', training package for learning the word-processing program. As I worked through it, I almost decided to return the whole apparatus to its manufacturer because of the tone of the dialogue that came up on the screen.

The training program told me what to do with the cursor. So I used the CURSOR command, and the screen gave me the condescending message:

Congratulations! You're moving the cursor and raring to go!

The assertion that I was 'raring to go' seemed an unwarranted assumption, but I ignored it, and went on to ERASE. I managed to use that command correctly. The machine said:

Wow! Am I impressed. Now let's see what you can do without any help.

I was struck by the artificiality of the written 'Wow!', and by the patronizing tone of 'Am I impressed'. I went on to the command EXCHANGE. I used it correctly, and received the comment:

You're hot! Quick! Try it again.

So I did, and I got it right again. The machine said:

Congratulations! Right again !!!!!!!

This is the over-fulsome tone of a primary-school teacher providing verbal rewards to encourage children to continue.

As if it could not believe that I had the wit, the machine greeted my correct operation of the MOVE command with:

Exactly! You're **really** sharp today. Now you can try it without any coaching.

I was beginning to be thoroughly irritated by the machine's patronizing tone. I tried COPY, and operated it correctly:

My, my, you do seem to be getting the hang of it! Now try doing it on your own.

I resented the sardonic tone of *My, my,* and the condescension of *you do seem to be getting the hang of it.* I could almost feel that primary teacher releasing my hand!

When I successfully used the WRITE command, the text-writer felt it appropriate to say to me:

Do you find doing things right all the time a little tiresome? Congratulations! You're doing great.

This is not only insulting, but also ungrammatical!

I did get things wrong occasionally (deliberately, of course, just to see how the program responded!). I got EXCHANGE wrong, and it said:

Well, you missed the boat that time. Why don't you try again.

I did it again and got it wrong again. It said:

Hey, pal! You missed again. Why don't you try it one more time.

The text-writer was presuming to call me his or her pal – certainly a misjudgement by this time!

I went on to LOCATE. I got LOCATE wrong, and the text-writer said:

Nope!!! That's not right. Maybe you should try that once more before you go on.

Nothing positive was gained by the slangy 'Nope!!!'. On the contrary, it stressed the effort the writer was having to make to be friends with me, and reinforced my negative response.

I did it again, and then went on to TAB. I got TAB wrong:

Have you ever felt like the ship was sailing without you?

Here was ungrammatical expression again. I began to be suspicious about the literacy of the writer who was addressing me through the screen.

I tried POINTER. I got POINTER wrong, and my interlocutor said:

If you can't get this one, you're looking out the window! Please go back and give it one more try.

Forced jocularity. That primary-school teacher again.

I went on to MOVE and got that wrong. It said:

> Have you been eating your lunch and not paying attention? Why don't you try again?

That American way of expressing an invitation – in this case, encouraging the operator to try again – using the 'Whydoncha' idiom, is not commonly heard by the user of British English. It is likely to be taken literally, and to produce a response such as 'Because I don't bloody well want to!'. Certainly, after so much irritation, it produced such a response from me.

I wondered if I was over-reacting. After all, the writer was from the USA, and the reader was British. Perhaps I was just being an unduly jaundiced reader. Perhaps I was just reflecting general British characteristics – stiff-necked, stiff-lipped, stuffy, and over-formal. Perhaps to an audience in the USA, the tone of that on-screen commentary would have been acceptable.

To find out whether I was alone in reacting against this on-screen text, I have watched the responses of 20 British 'naive users' as they have worked through the tutorial. Every one of them found the text offensively patronizing. To find out the response of readers in the USA, I have presented the text to audiences totalling more than 300 people at conferences of communication specialists in the USA. Those audiences were interested in tactics of communication, and might have been expected to appreciate sensitive and appropriate comments from the writer. They, too, found the text unsatisfactory.

How might the text have been improved? In general, by removing all apparent value-judgements, and all the attempts to be light-hearted. For example, when I operated COPY correctly, I was greeted with:

> My, my, you do seem to be getting the hang of it! Now try doing it on your own.

The first sentence evaluated my performance, and made a comment that seemed either sardonic or falsely hearty. I would have been happier to have received a simple *Correct* or *Right*, followed by a request such as *Now try using the command without prompts*.

In my experience, a comfortable, conversational, 'user-friendly' tone is best produced by the use of simple vocabulary in direct (second-person) address to the reader. I am well aware that there is a fine dividing line between expressing yourself simply and expressing yourself in an apparently patronizing tone. In an effort to make clear to readers exactly what they must do or understand, it is easy to slip over into apparent bossiness or condescension. But the tone of a text plays a substantial part in creating readers' responses to that text. Attention to tone is an important ingredient in the tactical mixture that creates good style.

Humour

This book is about style. A general debate about the introduction of humour, including cartoons, into technical documentation would therefore be out of place here. It is, however, legitimate for me to point out the effect on readers of attempts to write in a 'light-hearted' style by introducing quips, banter, and other forms of 'light relief' to the text.

The effect is likely to vary widely from reader to reader, and from culture to culture. Perhaps you feel that the comments I have made earlier in this section on the word-processing training text are unjustifiably harsh. If so, your reaction illustrates my point. And if you share my distaste for the style of that text, your reaction again supports my point: you and I presumably have different opinions from those of the writer of the text and the product manager who approved it. I suggest, therefore, that you should resist the temptation to be jocular unless you are absolutely confident that all your audience has the same sense of humour as you have.

Especially, resist the temptation to write down exclamations such as *My, my, Wow, Whoops, Hey,* and (again from my word-processing instructor) *BLAAAAAT!* or *You missed the boat that time.* Recognize that the acceptability of such comments differs in speech and in writing. If you are face-to-face with your interlocutor, then your body language, your cheerful voice, and your friendly smile all work to create the tone of encouragement you intend. Also, you can see from the body language (and perhaps hear from the curses) of your 'pupil' whether or not a jocular comment will be well received at that moment. The same comments, committed to paper or screen, may seem to carry a tone that is very different from the one you intended, or – even if the intended tone is detected – may be very inappropriate to the mood of your pupil at that moment.

Above all, recognize the difficulties created by attempts at humour, especially by attempts at a whimsical or ironic tone, for readers using English as a foreign language. Readers who do not have a very good command of English may well miss entirely the intended tone of many of the examples I have presented in this section. They are likely to try to interpret each word at its face value. Because they are expecting a technical discourse in the book or screen they are reading, they attempt to deduce some technical meaning from the sum of the words they read. Although native speakers of English should have no difficulty in recognizing the tone signalled by the exclamation mark in the following extract, readers using English as a foreign language may have difficulty with the idiom *which standard is which* and may not pick up the intended tone of *remarkably anonymous*:

> Place the three remaining microvials from Lesson 5 in their sleeves and screw a cap over each top. Keep a mental record of which standard is which; the microvials are remarkably anonymous in their sleeves!

Even professional translators would be in a quandary when faced with that text. They may be able to pick up the intended tone, but what should they do? Should they ignore the flavour of *remarkably anonymous*, and convert it into something like *the microvials all look the same*, or should they try to find an equivalent whimsical phrasing in their own language? Humour is difficult to translate; also, notions of what is humorous, and when it is appropriate to be humorous, vary from country to country, and from culture to culture. It is wise to write in a plain, simple, style, thereby reducing difficulties for readers and/or translators.

Large, international companies now spend considerable amounts of money on 'localizing' their documentation. Localizing means adjusting documents to make them appropriate to cultures and contexts different from those in which the originating writer(s) lived. By taking care not to create problems of tone, you may not only establish a more satisfactory rapport with your native-English readers, but also remove the need for expensive localization of your texts.

CHAPTER THREE

Avoiding 'distorted' English in computer-related texts

There is a trait in computer-industry writing that is not common elsewhere: it is the distortion of normal wording and phrasing. Consider these examples:

> If the call fails for any reason, the Gateway sends a 'Call Reject' message.

> Specify a mailbox to receive inbound X Net connects from the Y machine,
> The user confirms the clear by issuing...

None of these is normal English. In the first, the meaning was a 'call **rejected**' message. In the second, the normal English would have been 'X-net **connections** from the Y machine'. In the third, normal English would have been 'Confirms the **clearance**'.

One difficulty here is that words are used as unusual parts of speech. That by itself might be acceptable. But readers are troubled also by the ambiguity possible in *Call Reject*. Is that an imperative? Especially for users of English as a second language (to whom much technical documentation is now addressed), will the meaning be readily apparent?

Notice the same sort of ambiguity in the next three examples:

> If a status message 'transmit beginning' does not appear...

> ...provides requests that enable you to reference absolute locations in X and Y programs,

> ABC will not permit application of after-looks out of chronological order and will error if it should detect any...

In the first example, the meaning was: 'the transmission is beginning', not 'transmit the beginning again'.

In the second example, *to reference* is being used in a confusing way. *To reference*, I suggest, means to give bibliographical information about where you have obtained information. The meaning here is that it enables you to 'refer' to absolute locations. *To refer to* is not the same as *to reference*.

In the third example, the use of the word *error* is again both unusual and ambiguous. The word 'error' is normally used as a noun. For example, we discover an error, or make an error. The word we use as a verb to express the idea of making a mistake is to *err*. So we are surprised when we read that the ABC program *will error*. We wonder what it means. We try the word *err* in its place. Does that give the correct meaning – the program/machine will err? Is that right? Or did the writer mean one of the following possibilities:

- the program/machine will continue to function, but will function erroneously;
- the machine will function normally, but the program will cause an error in a calculation;
- the program/machine will present us with an error message.

Use of normal English, in combination with clearer thinking, would have removed confusion from readers' minds.

When writers write in this distorted way, it is not surprising that people outside the industry begin to say: 'Well, one of the problems with writers in the computer industry is that they are illiterate. They don't know how to write'. I do not believe it is true that writers in the computer industry do not **know** how to write in normal English. Equally, though, I do not believe there is any justification for these distortions.

There was a time when programmers were constantly short of space in computer storage and/or on screens; but modern computing capacity is such that restrictions on the length of texts or messages are rarely responsible for distorted English. In any case, shortage of memory space is no excuse for distortions in manuals and guides. Such documents are usually written by professional writers, and it is especially distressing to find a professional writer saying that a machine *will error*.

Distortion vs true jargon

The proper creation and use of jargon are not in question. To express new ideas, new words are needed: for example, *byte*, *diskette*, and *initialize*. Equally, I am not objecting to the adoption of existing words to convey new meanings within a computing context: for example, *gateway*, *bug* and *handshake*. The causes of concern are uses of English that distract and bewilder because they are unexpected substitutions for existing, adequate, familiar forms.

If a text is apparently written in English, the reader attempts to decode it in accordance with the normal conventions of English. Departures from normal language behaviour are at least jarring, and at worst incomprehensible.

The first time I read the sentences containing the following extracts, I had to pause to check that I was sure what they meant:

...record the length of the system stop...

...pausing until carrier detect is indicated by...

...a procedure known as Programmable Option Select...

Of course, I did not have to pause for long. It was fairly easy for me to work out what the normal wording would have been:

system stoppage	instead of	*system stop*
carrier detection	instead of	*carrier detect*
programmable Option Selection	instead of	*programmable Option Select*

But I had to do more than work out what the normal English would have been: I had also to wonder if a *system stop* was something different from a *system stoppage*; if *carrier detect* was something different from *carrier detection*; and if *Programmable Option Select* was something different from *Programmable Option Selection*.

Fortunately, I was able to ask the writers what they meant. All agreed that their unusual wording meant the same as the normal wording. But other readers would not have access to the writers, and might be forgiven for being at least uneasy about their interpretations, and at worst, badly confused. For readers using English as a foreign language, who are decoding in accordance with the normal conventions of English, such non-conventional wording is a considerable obstacle.

Perhaps you will be tempted to argue that the use of different wording in these computer-related texts is simply the creation of a special computing jargon. I dispute that. I have acknowledged already that, as in any new science or technology, new terms are needed to express new concepts or activities in computing. These terms can be entirely new coinages, like *byte* or *interface* (which is not listed in my 1973 edition of the *Shorter Oxford Dictionary*), or old words with special meanings in the new context, like *handshake* in *handshake protocol*, or *driver* in *device driver*. But the expression *system stop* is not needed to express a new concept or activity. The words *system stoppage* would have been entirely adequate. The expression *system stop* is a gratuitous departure from the normal conventions of English, which would have expressed the writer's meaning entirely adequately. My protest, then, is not against the creation of new terminology: it is against the creation of an unnecessary new version of English to surround that new terminology.

Reasons for keeping English 'natural'

There are four sound reasons why we should resist the temptation to create a new version of English for use in computer-related writing:

1. Natural forms of language are comfortable and unsurprising. They leave readers free to concentrate on the incoming message.

In contrast, distorted or unnatural forms are surprising; they distract our attention, even if only momentarily, from our main task of absorbing meaning from the overall text.

2. There are many **necessary** new terms that we must learn as we acquire specialist knowledge about computing. To make us learn a new version of 'general' English as well is to impose on us an unnecessary extra burden.

3. The distorted/variant language is often less clear than the ordinary/natural language it displaces.

4. Distorted/variant forms are major obstacles to readers using English as a foreign language, even to expert translators. Native speakers, with long experience of the ways in which English is used in varying contexts, may well be able to guess which normal word or structure is being displaced by a distorted variant. For speakers whose experience is less extensive, guess-work is not so easy.

This last reason is of great importance. Put yourself in the position of someone who is trying to interpret the following extract by converting it into his or her own language:

. . . creates a batch id with the edit status 'movable'.

Note, first, that you would need to have learned the special use in computing of the 'word' *id*. (That word, spoken as 'eye dee', is in general use in the USA, but is not much used in general discourse in Britain. It is not listed in the *Oxford Advanced Learner's Dictionary of Current English*.)

I suspect, though, that your main difficulty would be with the expression *the edit status*. In that phrase, a word that is normally used only as a verb, *edit*, is used as an adjective. If the meaning of the phrase *the edit status* was clear, the abnormal use of *edit* would be distracting but would not cause a breakdown in communication. But I, a native speaker of English, am not sure if that phrase is intended to mean 'the edited status', 'the editable status', 'the editing status', or something else. What hope has a reader for whom English is a second or third language?

Perhaps you will be tempted to reply that readers whose command of English is poor would have to have manuals translated into their own languages. But that would be merely to transfer the problem to someone else. Even expert translators would be in trouble with *the edit status*. They would have two choices: to guess which of the normal English words was intended, and convert that to the equivalent in their own languages; or to invent a distortion in their own languages. If the writer of the original extract had used normal English, we should all have been spared the uncertainty created by the distorted expression.

The following example adds another dimension to readers' problems, especially if they are using English as a foreign language:

> has included a protection capability in.... This key turns the protect on or off.

First, does *the protect* mean *the protection*? We guess that it probably does. But then another question arises: if *protect* is being used to express the notion of 'protection', what is *protection* being used for? Imagine that you have the task of translating that text. Would you be confident that you should use the same word in your new text to express the meanings of both *protect* and *protection*?

Normal language behaviour is commonly distorted in four main ways:

- new nouns are created, although adequate nouns exist already;
- word-forms that are customarily used as one part of speech are surprising and distractingly used as other parts of speech;
- normal word-order is distorted, and the abnormal word-orders introduced are not used consistently;
- inept compression creates confusing expression.

New nouns in place of adequate, existing nouns

In discussing *carrier detect*, and *Programmable Option Select*, I have provided already some examples of the creation of new nouns. In each of those cases, a word that normally functions as a verb is, surprisingly and unnecessarily, used as a noun, although an entirely adequate noun exists already. This habit of creating new nouns is probably the most common way in which normal language is distorted:

> Congestion control consists of single process, transmit management.
> (transmission)

> The panel also contains function select keys.
> (selection)

When distortion is combined with the common bad habit of pre-modifying with nouns, our task of disentangling meaning becomes even more difficult:

> In order to provide an XYZ network connect capability for the work-station,...

That extract, so the writer told me, was tangled and distorted English for:

To make possible a connection between the workstation and the XYZ network, . . .

Abnormal use of word-forms

Distortion is not restricted to the creation of strange nouns. Here is an assortment of surprising and unnecessary deviations from normal language:

...should be allocated via the ALLOCATE command. The process that does the allocate must be the same as . . .
(allocating/allocation)

Use double high, double wide characters to . . .
(height, width)

For operation security . . .
(operational)

Select the option QT at the main secretary menu . . .
(secretarial, secretary's, for secretaries)

Abnormal word-order

I have discussed extensively (section 2.8) the unusually frequent use in technical writing of nouns as pre-modifiers. That is one way in which technical writers depart from normal English word-order:

Two items of equipment are needed for the power subsystem maintenance.
(Two items of equipment are needed for maintenance of the power subsystem).

Another common form of distortion is reversed word-order. This is especially frequent in the creation of titles for activities (or 'functions'), and therefore is especially frequent in chapter headings and in on-screen menus. Reversed word-order is demonstrated by the following list of options in a menu:

Backbone build
Boundary update
Evaluate mode
Report selection
Exception reporting
NCP update

In normal English, when we think of an activity, the usual word to express it is a word ending in *-ing* (in grammatical terms a 'gerund', or verbal noun): *walking, typing, entering, building, selecting, installing, reporting, updating.* In computer-related writing, writers frequently introduce three distortions: one is the use of an imperative verb instead of an *-ing* word; the second is the reversal of normal word-order, the third is the omission of 'small' words.

In the menu of options in the previous extract, *backbone build* is a distortion of normal English: *build* is used where *building* would be normal; the word-order *backbone build* replaces the more normal *building (a/the) backbone* (or *building backbones*); the omission of *a* or *the* leaves us uncertain if the section will tell us how to build the sole backbone in the context, or how to build backbones in general.

The same three distortions are shown also by the second item in the list, *boundary update.* But then, in the third item, the writer (and/or the program developer?) adds to our confusion by apparently reverting partly to normal: *evaluate mode. Evaluate mode* seems to be the use of an imperative, *evaluate*, in place of the *-ing* form, *evaluating*; but at least the word-group seems to have been presented in normal order. The equivalent normal wording could be *evaluating a/the mode* (or *evaluating modes*).

But because I have read so much distorted, computer-related documentation, I am still uneasy. Is there another common distortion here? Are we going to read about *the evaluate mode*? Is *evaluate mode* a similar distortion to *the edit status* that I discussed earlier?

The fourth item on the list is utterly confusing. If you had to translate *Report selection* would you convert it to the equivalent of *reporting selections* or *selecting reports*?

Presumably, the fifth and sixth items would be *reporting exceptions* and *updating (the) NCP* in normal English.

I can see no valid reasons for these distortions. They are not true jargon: they are not new words or turns of phrase that are needed because existing language is inadequate; they are unnecessary departures from normal wording. They disturb native speakers of English, and cause considerable confusion for translators. If natural English had been used, there would have been no cause of concern for anyone:

Building the backbone	*or*	Building backbones
Updating the boundary	*or*	Updating boundaries
Evaluating the mode	*or*	Evaluating modes
Selecting the report	*or*	Selecting reports
Reporting the exceptions	*or*	Reporting exceptions
Updating the NCP	*or*	Updating NCPs

In discussions of this topic with people who live in the world of computing, I have been told that they 'get used to' this distorted version of English; they are so accustomed to it that they no longer notice that they are having to sort out a jumble of different structures.

My answer is that they are having to work harder than is necessary. Though they may not be conscious of it, the extra effort they are having to make is probably a cause of the widespread lack of enthusiasm (near-universal hostility?) for computer manuals. They would not find texts written in normal English **harder** to understand. Indeed, when readers are asked to choose from several styles (as in my surveys, discussed in Chapter 1 and the Appendices), the preference of the majority is for direct, natural language. Add to that the fact that natural language saves an immense amount of time and effort for translators and other readers for whom English is a foreign language, and we have a powerful argument for avoiding distortion (or 'peculiar English') wherever possible.

Confusing compression

In discussing the list of menu options (*backbone build*, . . .) in the previous section, I have given some examples of the confusion caused by 'compressed' style – the omission of so-called small words like *the, a, an, all, some,* and *it*. Sometimes the omissions create statements that are sufficiently silly to be quite clear. When you read the following extract, it becomes obvious immediately that the writer did not intend **you** to return to HQ in a padded envelope:

Remove the old disk. Return within seven days in padded envelope to HQ.

But the inclusion of *it* (*return it*. . .) would have removed all ambiguity.

Similarly, I was amused when I first read this caution at the beginning of an instruction sheet for handling cartridges for a printer:

Do not stand on end or turn upside down.

Of course, I could supply the missing words:

Do not stand the cartridge on its end or turn it upside down.

but more careful writing would have removed the ambiguity.

The translation of that instruction sheet into French demonstrated one way in which translators cope with ambiguities – by perpetuating them in the translation:

Ne pas mettre debout ou à l'envers

More disturbing is an example like this:

Voltage values are seen through small windows in panel. Switch ranges from 100 to 240 in six steps, and is positioned by turning...

When I read that text for the first time, I took *switch* to be an imperative verb. It was not until I arrived at *and is positioned...* that I realised that I had been misled by omission of *the*: *The switch ranges from....*

Omission of 'small' words is particularly common in instructions:

Delete data from XYZ files.

No doubt this type of instruction is produced because writers want to be concise; but conciseness should not be gained at the cost of clarity. Does the instruction above mean 'delete all the data' or 'delete some of the data'? Do we delete data from all the XYZ files or just from some of them? Sometimes, the context is such that there is no ambiguity; but frequently the meaning is not clear. Writers rely on readers supplying the correct interpretation. Once again, this often presents special difficulties for translators. I recommend that you resist the temptation to use a compressed style. Develop the habit of writing fully at all times.

Need for advice on style

Developers and programmers, with their minds full of filenames and commands, can perhaps be excused for lapsing occasionally into language that reads more like Fortran than English. Most computer specialists would not claim to be skilled communicators. Many have little sensitivity to the clumsiness and inaccuracy of the texts they produce. All too often, though, the distortions they create are imitated and carried over by professional technical writers into manuals and on-screen texts. For those writers, life in the computer industry seems to be having a serious desensitizing effect.

If the computer industry wants to create a private code, understood only by insiders, it is welcome to do so; but if it wants to present its information clearly and attractively to the world at large, using the code that we call English, it must conform to the conventions of that code.

Let me repeat that my attack on distorted English is not an attack on the use of true jargon. Nor is it an attack on programming language. The examples I have shown in this section are not examples of the coding that is designed to make computers function. They are examples from what are supposed to be the natural-language presentation of information by manufacturers of computing equipment to their customers. The peculiarity and the uncomfortable qualities of much in these presentations emphasize the need for us to make sure in future that system designers, programmers and writers co-operate and produce **together** the technical information that is to go into the manuals and the machines. The language used in the interface between manufacturers and users must be natural and clear, not distorted and confusing. System designers and programmers, not just writers, must be trained to communicate clearly in good, natural style.

CHAPTER FOUR

Style for instructions and procedures

Operating instructions are written so that a process, a procedure, a plant or a piece of equipment can be operated unhesitatingly, accurately and economically. The instructions must therefore be:

accurate:	the exact meaning of each instruction must be clear;
comprehensible:	each instruction must be manageable, not too involved or overloaded with information, not blurred by explanation and description;
adequate:	there must be enough information to permit operation, or at least there must be exit branches specifying where additional information may be found;
complete:	all feedback loops and branches must be specified;
in sequence:	the steps must be in proper order;
safe:	warnings must be prominent and well placed;
acceptable:	the tone must help gain the response desired; it must not produce resentment or hostility.

To meet these requirements, it is necessary to go beyond consideration of style. We must consider the selection, arrangement and physical presentation of information; we must have in mind constantly the frames of reference of the receivers, their levels of expertise, their past training (if any), and the contexts in which the instructions have to be used.

A full discussion of tactics for organization and layout of instructions is beyond our scope here; but consideration of the following example will make plain how features of style are inter-related with features of selection and organization:

Tubular unit operating instructions
1. . . . . .
2. . . . . .
3. . . . . .
4. . . . . .
5. The furnace interior should be inspected every four hours for insufficient burner operation. Look for ragged flame, smoke or 'points of flame'.
6. Check the stripping column bottoms level controller AB12. Raise the set-point on AB12 to . . . X . . . Level pen should follow. Float freedom can be tested by means of the rod in the base.
7. Twice a shift check the whole system for process leaks. Take particular care in the furnace and pre-heater area.

Does the first sentence in Instruction 5 give advice or an instruction? *The furnace interior* should *be inspected* seems to leave an option open. The statement should be imperative: *Inspect* . . . .

We are told to look for *ragged flame, smoke or 'points of flame'*; but we are not told what to do if we find some! A common weakness of instructions is inadequacy of information, especially omission of information about what to do if things are not as they should be.

We are told to check the stripping column bottoms level controller AB12. What does this mean we have to do – just look at it? In what circumstances are we to do something to the controller, and in what circumstances do we leave it as it is?

Presumably, raising the set-point to...X...will cause no difficulty. But does it matter if level pen does **not** follow? And what should we do if it does not? Inadequate information again, and *should* makes the achievement of level pen seem optional.

The final sentence in Instruction 6, *float freedom can be tested by means of the rod in the base,* is not an instruction: it is a comment. What relevance does it have to the instructions? When should float freedom be tested? Should the float be free? What should we do if it is not? It is not much help to us to tell us that something *can be* done. Such wording makes a comment, not an instruction.

Twice in each shift we are to check the whole system for process leaks. If we are on an eight-hour shift beginning at 8 a.m., and it takes 20 minutes to walk around the system, is it satisfactory for us to walk round once from 0800 to 0820 and again from 0830 to 0850? And what are we to do it we find any leaks? Merely write a note in the log book? Shut down the plant? Run? We are told to take particular care in the furnace and pre-heater area. Why? Is it slippery in there?

Plainly, the adequacy of these instructions would depend greatly on the expertise and previous training of the users. But even if the users had been given a detailed training course, there should have been some cross-reference in these instructions to relevant parts of training manuals or other explanatory documents; and many of the ambiguities should have been removed by more careful choice of language and style.

Vague words

The danger of a vague word like *should* stands out obviously. Each use of it in any instructions should be scrutinized closely, as it may reduce to a comment an item intended as an instruction. If it is necessary for a reactor to be warmed for two hours before use, to write *the reactor should be warmed* will not be adequate. You will have no grounds for complaint if the warming is omitted, as you have used an optional form: you must write *warm the reactor.*

The word 'will' can similarly reduce an instruction to an observation.

The filter will be changed every six hours simply states a fact: no-one is instructed to change the filter. To make this an instruction, you must use an expression such as *change the filter*.

Another vague word is *check*. Again, you will have inadequate grounds for complaint if nothing happens in response to the instruction *check that the drain valve to the top of the column is closed*. Is the operator simply to report the state in which he or she finds the valve? If you want the valve closed, you must write *ensure that the drain valve to the top of the column is closed*.

Other dangerous words are:

select: *select first gear*: does this mean there is a choice of which gear is to be used first?

locate: *locate the retaining knob on the baseplate*: does this mean 'find' or 'connect'?

replace: *replace the pin that is damaged*: does this mean 'put it back' or 'fit a new one'?

Definitions

To be safe, it is necessary to define service activities and process activities clearly. Probably, the best place to do this is in a training programme, with clear definitions written in a training manual; but if necessary, definitions should be given at the start of each document, as is attempted in this example from a rescue harness servicing procedure [21]:

SERVICING NOTES

1. **Glossary**   The servicing operations detailed in this bay servicing schedule have the meaning given in the Concise Oxford Dictionary except for the following:

   a. Inspect   Review the work carried out by tradesmen to ensure it has been performed satisfactorily.

   b. Check   Make a comparison of a measurement of time, pressure, temperature, resistance, dimension or other quantity with a known figure for that measurement.

   c. Test   Ascertain, by using the appropriate test equipment, that a component or system functions correctly.

   d. Examine   Carry out a survey of the condition of an item. For example, the condition of an item can be impaired by one or more of the following:

(1) insecurity of attachment;

(2) cracks or fractures;

(3) corrosion, contamination or deterioration;

(4) distortion;

(5) loose or missing rivets;

(6) chafing, fraying, scoring or wear;

(7) faulty or broken locking devices;

(8) loose clips, or packing, obstruction of, or leaks from pipelines;

(9) discoloration due to overheating or leaking of fluids.

e. Operate — Ensure an item or system functions correctly, as far as can be ascertained without the use of test equipment or reference to measurements.

f. Replenish — Refill tank, bottle, or other container, to a pre-determined level, pressure or quantity, and where necessary:

(1) remove caps, or covers from filler orifices and/or drains;

(2) clear orifices;

(3) fill container as directed in item operation;

(4) ensure drains are free from obstruction;

(5) ensure gaskets and caps or covers are free from damage;

(6) refit caps or covers;

(7) fit locking devices as necessary.

g. Fit — Correctly attach one item to another.

h. Refit — Fit an item which has previously been removed.

j. Replace — Remove an item and fit new or serviced item.

k. Disconnect — Uncouple or detach cables, pipelines or controls.

l. Reconnect — Reverse of sub-para k.

Each word or phrase in an instruction-sheet must be considered in relation to the readers' background knowledge. Will they understand expressions such as:

Slacken the cheesehead retaining screw.

Remove the camerated segment.

Examine the wheel disc for signs of fretting.

Slide the guideplate right home.

Pull smartly upwards.

The first three examples use specialist jargon; will all readers understand it? If so, by all means use it. If not, either explain it, or avoid it altogether. The last two examples make use of images familiar to native speakers of English: but will they be comprehensible to readers for whom English is a second language? (Even in the USA, the idiom *right home* is not in common use, and would cause difficulty for many readers.)

Constant attention to the readers' frames of reference is vital. Do they know what is meant by *connect up the supply tank and the liquid gas vaporizer **in the usual way***? Do they have in mind the same ranges as you have when they read your instruction to *ensure there is **little or no** pressure in the reactor* or *set the pressure in the main to **about** 40 p.s.i.*?

And will it be clear who is to act in each instruction? Clear indication of the agent, if it is necessary to specify one, will depend on two things: the format of the complete set of instructions and the style.

Imperatives

If you specify at the beginning of a set of instructions the readers for whom you are writing, you can write freely in impersonal style. For example:

Instructions for the bagging machine operator
The filter is to be changed every six hours.
The filter must be changed every six hours.

Strictly, these two expressions are descriptions, not instructions. Certainly, given the title above them, they express an obligation; but they are not imperatives, and it would be wiser to write:

Change the filter every six hours.

If you have not specified for whom those instructions are intended, you may find the change not made 'because it didn't say it was my job to do it!'

Occasionally it is desirable to specify an exact agent:

The supervisor/Mr Brown is to change the filter.
The supervisor/Mr Brown must change the filter.

In these circumstances, too, you may find that the change has not been made 'because it says it's Brown's job, not mine!' You can guard against this risk by ensuring that your work organization has specified who is to act in Brown's absence.

For clarity and acceptability, write instructions as a sequence of steps, each beginning with the imperative form of a functional verb (*raise, lift, turn, switch, ensure, connect*). Occasionally, siting or conditional phrases can appear before the verb:

At panel B, switch...
If the reactor is partly full, connect...

Occasionally, for variety or emphasis, it may be useful to turn to the impersonal style to specify an agent, using the *must* or infinitive forms; but a crisp sequence of short, imperative instructions will normally produce best results.

Manageability and sequence

Two final points, about manageability and sequence, can be made with one more example:

Flanged valves: test procedure
1. Ensure cleanliness as described under 'Preparation' paragraph 15.
2. If a blow down ring is fitted, remove the blow down ring lock screw and then turn the tooth towards the right (counter-clockwise viewed from above) thus raising the blow down ring until it meets the disc holder, then lower by turning the tooth towards the left (clockwise viewed from above) to half the number of teeth shown in either table A or B, (depending on service of valve), but not less than two teeth. This may be done by reaching in through the locking screw hole with a screwdriver. After the blow down ring is relocated, replace the lock screw and tighten; make sure the lock screw rests BETWEEN the teeth so as not to distort the ring. Side pressure on the blow down ring can cause leakage. The blow down ring should have some play after the screw has been tightened, but should not be free to rotate.

   NEVER RAISE THE RING WHILE PRESSURE IS ON THE VALVE, AS THE VALVE MAY 'POP' AND COULD CAUSE INJURY.

The first point illustrated here is the clumsiness of writing that attempts to amalgamate several instructions into a single long paragraph. These instructions would have been much clearer as a sequence of short, imperative instructions.

The second, **very** important point, is the position of the warning. After telling us at the top of the paragraph to raise the blow-down ring, the writer warns us at the end that to do so while pressure is on the valve may be dangerous. Too late!

I am constantly astonished by the frequency with which warnings are given **after** dangerous steps in instructions. The correct place for warnings is **before** the dangerous step. And warnings should be clearly separated, not buried in the text, as in:

The dilution of sulphuric acid should preferably be carried out in clean earthen-ware or lead-lined tanks which should be about half filled with pure water, and a suitable quantity of concentrated acid poured slowly into it, the mixture being stirred all the time with a clean piece of wood. Concentrated acid is much heavier than water, and will tend to settle at the bottom. On no account should the water be added to concentrated acid, as the water would remain on the surface and the intense heat generated would cause it to boil and splash the corrosive liquid about. The effect might be so violent as to be almost explosive.

CHAPTER FIVE

Style for descriptive and explanatory writing

5.1 DESCRIPTIONS AND EXPLANATIONS

Most handbooks and other forms of technical documentation consist of description, explanation and instructions. They contain little theoretical analysis, evaluation, argument, or expression of opinion. The most suitable style for most handbooks and technical documents is therefore likely to be impersonal. There is little natural call for the personal constructions suited to debate and evaluation. But to say the style should be impersonal is not to say it should be ponderous and dull. To write impersonally does not necessarily require use of a passive, roundabout style:

Not: In so far as the carriage of current is concerned this is achieved through the medium of bus-bars.

Not: The current is carried by means of bus-bars.

But: Bus-bars carry the current.

Voice

Expository writing can be clear, lean and vigorous. For example, compare and contrast the two versions of a description below. The 'better' version is lucid and readable. Its sentences are of varied length, but on average they are short and not complex. Verbs are mainly active, and the vocabulary is 'plain' but accurate. It illustrates the cumulative benefits of writing directly and actively:

Not: The extrusion or capillary rheometer, which is capable of recording a high shear-rate range, and may be utilized to obtain a measure of normal stress, is generally considered to be the most suitable apparatus for the determination of the characteristic flow behaviour of unvulcanized silicone rubbers. In this apparatus, extrusion of the material under consideration is carried out through a cylindrical die. The apparatus consists essentially of a cylinder in which the material is placed and a ram by which the rubber is forced along the cylinder, being extruded through a die at the end of the cylinder. The apparatus is attached to a tensometer and measurements of the force required to force the material through a die at different rates may be determined. A curve relating the volume extrusion rate and the back pressure may be produced without difficulty by the operator.

Better: The extrusion or capillary rheometer is the best for finding
out how unvulcanized silicone rubbers flow. Its shear-rate
range is high, and it can be used to measure normal stress.

In this rheometer, a ram forces the rubber through a cylinder and out through a die at the end of the cylinder. A
tensometer attached to the apparatus measures the force
required to push the rubber through a die at different rates.
The operator can easily produce a curve relating the volume
extrusion rate and the back pressure.

It is essential to remember, though, that active-voice constructions
and passive-voice constructions are not lightly interchangeable (this
point was discussed in section 2.11). Consider the following two versions of a description of an item of equipment, a microwave drier for
drying granules of a compound from which pharmaceutical tablets will
be made. Version 1, in the left-hand column, was the original; Version
2, in the right-hand column, is a revision into a style that uses more
active verbs. Do the two versions give precisely the same emphases?

Version 1
The XXX prototype dryer is
illustrated in Figure 1.
Microwave energy is beamed
into the microwave housing
by twelve magnetrons,
each mounted on its own
waveguide. Uniform
distribution of microwaves
through the polypropylene
window into the drying
chamber is ensured by a mode
stirrer. The temperature of the
granules is monitored by a
temperature probe, which
detects any overheating. The
E-field (electric field) is
monitored during the drying
cycle by a detector mounted on
the dryer chamber. (A measure
of the excess energy in the
dryer chamber is provided by
the E-field: an increase in the
excess energy is experienced as

Version 2
The XXX prototype dryer is
illustrated in Figure 1. Twelve
magnetrons, each mounted on
its own waveguide, beam
microwave energy into the
microwave housing. A mode
stirrer ensures uniform
distribution of microwaves
through the polypropylene
window into the drying
chamber. A temperature probe
monitors the temperature of
the granules, and detects any
overheating. A detector,
mounted on the dryer
chamber, monitors the E-field
(electric field) during the
drying cycle. (The E-field
provides a measure of the
excess energy in the dryer
chamber: the excess energy
increases as the batch dries.)
The E-field signal passes to the

the batch is dried.) The E-field signal is passed to the controller. The controller is set by the operator such that the magnetrons are switched off when a level that indicates suitable dryness for each granule formulation is reached by the E-field.

controller. The operator sets the controller to switch off the magnetrons when the E-field reaches the level that indicates suitable dryness for each granule formulation.

As we saw in section 2.11, the use of passive constructions moves the emphasis of a sentence. In version 1, the emphasis in sentence three falls on *uniform distribution*; in version 2, the emphasis in sentence three is on *a mode stirrer*. Overall, version 1 gives more emphasis to what happens; version 2 gives more emphasis to how it is done. Without doubt, your writing will seem more direct and energetic if you take every opportunity to write actively rather than passively; but as the two versions above show, the verb-forms you should choose depend on where you want your emphasis to lie.

The two versions above are part of a description of an 'object'. Usually, when we describe an object or an idea, it is possible to choose between active and passive styles. When we describe processes or procedures, it is frequently difficult to avoid writing passively.

Here is a description of how the same microwave drier is operated:

A charge of wet granules is placed in the dryer bowl. The bowl is eased off its transporting trolley on to the support table. The table and bowl are raised into the drying chamber by operation of three pneumatic cylinders. The vacuum pump is then started, and the chamber is evacuated through the vacuum take-off line. The valve in the air-purge inlet is opened and adjusted, to set up a controlled air-flow. (The air-flow is needed to carry the water vapour away from the dryer, and to prevent it re-condensing on the dryer walls or on the granules.) The microwave power is then switched on.

I know many editors (and teachers of technical writing) who would reach automatically for their correcting pencils at the sight of all the passive constructions in that description. But what happens if we try to make the whole text active?

To make it active, we have to supply an agent to do the work. One way is to use *The operator* . . . :

The operator places. . . The operator eases. . . The operator operates
. . . in order to raise. . . The operator starts. . .

Repetition of *the operator* becomes very tedious. To relieve the tedious repetition, we could replace some uses of *the operator* with a pronoun. But then we run up difficulties with sexism, or we begin to produce clumsy repetition of he/she:

The operator places...He/she eases...He/she operates...He/she starts...

One way to avoid sexism is to make everything plural; but that seems very unnatural, and seems to suggest that several operators are needed for the procedure:

Operators place...They ease...They operate...They start...

Another way to avoid sexism is to use *you* or *one*. Certainly, the use of *you* permits a very direct style; but excessive repetition of *you* soon becomes tedious, and there is no pronoun with which to introduce variation. The use of *one* seems unacceptably arch in English (its tone is very different from the use of *on* in French), and frequent use of *one* again becomes tediously repetitious.

It becomes plain, therefore, that a sequence of passive constructions may often be the least objectionable style to use for describing processes and procedures.

However, there is one way of presenting a description of a process, the 'playscript' layout, that enables you to write in a simple sequence of active constructions. The 'playscript' lay-out, described in detail by Leslie Matthies in *The New Playscript Procedure* [22], usually has two columns, one headed 'Responsibility' and the other headed 'Action':

Responsibility	*Action*
QA Inspector and Production Manager **jointly**	Decide on the acceptability of the samples.
QA Inspector	Places 'Unacceptable' stickers on all containers of rejected material.
	Stores all containers of rejected material in Area X of the warehouse.
	Stores acceptable material in...

If, however, there is only one 'actor' with a responsibility, it is possible to put the actor's title at the top of a text, and present the text in a single block. On such a lay-out, our description of how the microwave drier is operated would look something like this:

To operate the drier, the operator:

Places a charge of wet granules in the dryer bowl.

Eases the bowl off its transporting trolley on to the support table.
Operates pneumatic pumps 1, 2, and 3.
(The pumps raise the table and bowl into the drying chamber).

Starts the vacuum pump.
(The pump evacuates the chamber through the vacuum take-off line).

Opens and adjusts the valve in the air-purge inlet.
(Opening the air-purge inlet sets up a controlled air flow. The air-flow is
needed to carry the water vapour away from the dryer, and to prevent it
re-condensing on the dryer walls or on the granules.)

Switches on the microwave power.

This playscript lay-out is excellent for job descriptions, and for presenting quality-assurance procedures. It is not usually appropriate for
descriptions of computing processes in guides or on-screen presentations; in those texts, writers most frequently need to give step-by-step
instructions, not step-by-step descriptions; but occasionally, for technical brochures that precede operating documents in the equipment-
selling sequence, playscript can be very effective.

Tense and mood

When you are writing descriptions, it is important to use tenses and
moods consistently. Focus carefully on whether you are describing
what exists/happens, what will exist/happen, or what would exist/
happen. Avoid swinging between present and future:

Don't write   Chip-selection logic uses A, B, and C to select a byte or
word. If the X is being addressed, both chip-selection
lines will be inhibited. At the end of the memory cycle,
Y is deasserted. On the next negative 12 MHz clock, the
request latch and the RAM-state counter will be reset,
and the arbitrator will continue to scan for requests.

Prefer   Chip-selection logic uses A, B, and C to select a byte or
word. If the X is being addressed, both chip-selection
lines are inhibited. At the end of the memory cycle, Y is
deasserted. On the next negative 12 MHz clock, the
request latch and the RAM-state counter are reset, and
the arbitrator continues to scan for requests.

It is often difficult to keep clear in the reader's mind the difference between features or possibilities that are hypothetical or possible, and those that **do** exist in the equipment, process or system.

The clarity with which this distinction is made depends greatly on careful control of the sequence of verb forms. If a statement begins with a condition, the verbs that follow must be mainly conditional forms. The sequence for discussing a fact or condition that **could** or **might** exist now or in the future should be:

If X **were** true (at the moment, or in the future), Y **could/would/might** happen, and Z **could/would/might** be possible.

The sequence for stating that, though we have no supporting evidence, we are accepting for the moment that a fact or condition **does** exist should be:

If X **is** true (now), Y **can/will/may/could** happen and Z **is/can/will/may be/could be** possible.

The sequence for discussing one or more possibilities **in the future** should be:

If X **is/becomes** true (at some time in the future), Y **may/will/could** happen and Z **may be/will be/could be** possible.

The sequence for expressing a fact or condition that **could have/ may have** existed at a specified time in the past, though we do not know whether or not it **did** exist should be:

If X **was** true (at that time), Y **may have/could have happened/been happening** and Z **may have been/could have been** possible.

The sequence for expressing a fact or condition that could have been true, but was not, at a specified time in the past should be:

If X **had been** true (at that time), Y **would/could/might have happened** and Z **would/could/might have been** possible.

Unfortunately, writers frequently confuse what **is** true with what **might** or **could** or **should be** or **will be** true:

In order to carry out a major examination on any one transformer the following procedure should be followed. The 11 000 volt 250 MVA OCB in the Engine House should be isolated and drawn out. A notice should be placed on the switch to give warning of work being carried out on the system. The 33 000 volt 1000 MVA OCB is then tripped, either from the Engine House or from the outdoor substation. The control panel door on the 1000 MVA OCB was then opened and the Control Key removed, this would then trip the 1000 MVA OCB if the above methods were not followed. The Control Key is then used to unlock the mechanism on the pole isolator. After isolation the mechanism is again locked using the Control Key. The system should then be earthed to drain off any remaining charge.

As a reader, I was unable (even though I had the help of surrounding information) to make out which meaning the writer wanted me to take from the following part of his report:

The simple index might not have been valid in this case because the lateral spacings are also changed and this could affect the magnitude/sharpness relationship explained in the notes.

Was he talking specifically about the implications of one case he had just described:

The simple index possibly was not valid in this case because the lateral spacings also were changed and this would have affected the magnitude/sharpness relationship explained in the notes.

Or was he pausing to reflect on the general implications of his findings:

The simple index may not be valid in cases such as this because the lateral spacings also are changed and this could affect the magnitude/sharpness relationship explained in the notes.

Lay-out

In the discussion of sentence structures (section 2.2), I warned against stringing together too many short sentences. But in descriptive writing, occasional use of a terser, smaller-step style of organization and expression can be helpful to readers. Such a style would seem 'bitty' and irritating if it were used for long stretches of writing; it would certainly be unsatisfactory as a style for sustaining an argument or discussion in a report; but readers who are following an account of a complex process, especially those who are trying to **learn**, not just to understand, the information being presented, often find it helpful if parts of the text are presented in small, easily digestible steps.

Compare the following passages. The first version is in cumbersome 'traditional', passive, involved style. The second version is more direct and readable. The third version demonstrates how this type of information can be stepped and emphasized for particular purposes (such as commentary printed beside a sequence of small illustrations or diagrams):

Version 1

Completion of Supertron triggering is followed by commencement of anode run-down and coupling back of the negative going waveform to the grid through V54A and a timing capacitor selected by Sw9/5, causing a 'Miller' type run-down, at a speed controlled by the selected timing capacitor and RV51 setting. This run-down, 'caught' at 106 volts by V52B, preventing possible jitter when the valve 'bottoms', is followed by regenerative stable state resumption by the Supertron, causing a sudden screen potential fall differentiated by C185 and R378, and triggering of the time-base.

Version 2

After the Supertron has been triggered, the anode begins to run down and the negative-going waveform is coupled back to the grid through V54A and a timing capacitor (selected by Sw9/5). This gives a 'Miller' type run-down. The speed of the run-down is controlled by the timing capacitor selected and by the setting of RV51. The run-down is 'caught' at 106 volts by V52B, preventing jitter which might otherwise occur when the valve 'bottoms'. At the end of the run-down, the Supertron regeneratively resumes its stable state; this causes a sudden fall in the screen potential, which is differentiated by C185 and R378, and triggers the time-base.

Version 3

After triggering of the Supertron, the anode begins to run down.

The negative going waveform is coupled back to the grid through V54A and a timing capacitor. The timing capacitor is selected by Sw9/5. This gives a 'Miller' type run-down.

The speed of the run-down is controlled by:

(a) the timing capacitor selected;
(b) the setting of RV51.

The run-down is 'caught at 106 volts by V52B. This prevents jitter which might otherwise occur when the valve 'bottoms'.

At the end of the run-down, the Supertron regeneratively resumes its stable state.

The resumption of this stable state causes a sudden fall in the screen potential, which is differentiated by C185 and R378. It also triggers the time-base.

But let me emphasize that the style in the third version should be used sparingly – for **parts** of texts only. Readers react strongly against the 'childishness' of whole books presented in this style. To present a mass of information that needs to be broken into many small steps, consider using integrated verbal and graphical presentation techniques, such as flow-charts and algorithms, to bring order, clarity, and variety to the presentation.

5.2 SPECIFICATIONS

Specifications must, above all, be unambiguous. This requires writers to be explicit about who is to do what, where, when, in what order, how, and with what materials. These are mainly matters of content: but style also needs care. An ill-chosen word can leave a gap through which much time, money and goodwill can leak away. The following extracts from specifications exemplify the most common weaknesses that are matters of style:

Kiln instrumentation and control specifications
  1. Each pair of kilns will be provided with the following control equipment:...
  4. The instrumentation and control scheme shall be designed and...
  6. XY indicators shall match the controllers and recorders, if possible.
  12. ...must be capable of withstanding an overload of approximately 10x range.
  14. Pneumatic supplies are required for cylinder actuators.
  29. Scan-fold chart design is preferred.
  42. Prices should be shown ex factory.
  52. It would be advisable for X Co (the contractors receiving the specifications) to have insurances in force to cover liability for...

Item 1 introduces one of the commonest weaknesses in the writing of specifications – the use of *will*. Unfortunately, in casual day-to-day usage in British English, we use *will* in three ways:

1. as the third person form of the future tense of the verb *to be*, to make a simple statement of something that will happen in the future:
   'The reactor will be built in two stages.'
   'The frame will be constructed in 3 in tubular steel.'
2. to express determination:
   'We will build the reactor, in spite of local opposition.'
   'I will construct the frame in 3 in tubular steel, although you disapprove.'
3. to express obligation:
   'The contractor will ensure that the reactor is complete by...'
   'The frame will be constructed in 3 in tubular steel.'

None of these uses is 'wrong'. In general, a word's 'meaning' is what most people use it to mean, and even though a word is frequently used in several different ways, there is rarely any ambiguity or misunderstanding, provided the context makes clear the speaker's intention. But the differences in meaning between uses 1, 2, and 3 above are signalled mainly by intonation and context. In specifications, we cannot leave so much to chance. As far as possible, we must use words in a way that leaves no room for varieties of interpretation; and at points where we want our text to have mandatory force, we must choose words with particular care.

In British contexts, *will* used in third-person constructions is not regularly recognized as having mandatory force. It would be possible for contractors to argue that they understood Item 1 to imply that someone else was to provide the kilns with the control equipment.

The word *shall* is more reliable. Although *shall*, too, is used in a variety of ways, in British contexts *shall* is usually accepted as having mandatory force in third-person constructions. Though Item 1 might be challenged, the obligation intended by Item 4 would not normally be open to question.

However, writers of specifications would be wise to avoid all possibilities of confusion by avoiding both *will* and *shall* as means of expressing obligation and liability. It is possible to do so by using *must* or *is/are to be* instead.

Items 6 and 29 demonstrate two common weaknesses in specifications: the addition of the words *if possible* and the use of the expression *prefer(red)*. Writers of specifications should make up their minds whether or not features or qualities are essential. There can be no complaint if expressed preferences are not met. If certain features or qualities are essential, there should be no option left open; if they are not essential, there is no value in mentioning them in the main specification. Of course, it may be true that the features or qualities are not vital but are desirable. In those circumstances, it is preferable to state that fact outside the main specification, or at least in notes set in a distinctive way, separate from the obligatory items of the main specification.

Item 6 is typical in another way, too: in its careless use of the word *match*. Does this imply equivalence in size, shape, manufacturer, material, colour? Words like *similar*, *like*, *matching*, and *equivalent* are all dangerous when used in specifications.

Item 12 illustrates a trap that writers of specifications and of operating instructions often stumble into: inexact specification of quantities or values. What variation is permitted by the phrase: *approximately 10× range*? Plus or minus 0.5? Plus or minus 1.0? Plus or minus 1.5? Almost always, a safer way of indicating quantities of values is to specify *not more than*, *not less than* or *between*.

Item 14 does not specify. The form *are required* is as weak as *will*, as it simply states or comments on a need. The item should read: *Pneumatic supplies **must be/are to be** provided for the cylinder actuators.*

Finally, Items 42 and 52 both apparently leave options open again. If matters are optional, they should be distinguished clearly from obligations. Writers may argue that the use of *should/would* or *preferably* does distinguish options from obligations: I accept that argument in theory, but would contend that in practice it is wiser to separate the listing of options and obligations, either by using separate documents, or by having distinctive lay-out within one document. By allowing themselves to use *should* and *would*, writers often carelessly express items as options when they really want them as obligations.

CHAPTER SIX

Style for correspondence

In most correspondence, technical writers are concerned with passing on administrative decisions or test results, with reporting new findings, with answering customers' enquiries, with following-up contacts that have been made by Sales Departments, or with discussing matters relating to materials and machines they possess or want. The aim is usually to transmit information exactly or to make a precise enquiry.

There are, of course, occasions on which it is necessary to write a letter of apology, explanation, or conciliation; it may be necessary to produce letters of complaint, letters presenting bad news, letters reminding friends and enemies of obligations, letters passing on congratulations, and letters of commiseration.

There is also widespread activity in what might be described as sales writing – the use of letters to circulate information about products and processes. But these types of correspondence are not the normal day-to-day concern of scientists and engineers, so I propose to confine my discussion of letter-writing to the comparatively straightforward tasks of giving or seeking technical information.

The aim of scientific correspondence is normally to transmit information directly and specifically: accordingly, writing should be direct and specific, and much of the advice given earlier in this book applies again here. But there is an additional factor to bear in mind. That is, that letters help to create 'the company image' of organizations. Everything we write not only passes information but also creates an impression of ourselves in the readers' minds. High-flown style creates an impression of pomposity; roundabout phrasing creates an impression of tortuous thinking; impersonal, passive writing creates an impression of inert facelessness. A business letter is written on behalf of a company or organization: inevitably, the character that suffuses the writing is partly transferred to the organization.

'Correspondence English'

Do you want to give the impression of a cold, ponderous juggernaut, grinding impersonally and unswervingly along immutable tracks? If so, write something like this:

> Receipt of your letter reference AB/CD/1234 dated 14th inst. is acknowledged with thanks.

> It is regretted that this Company is unable to comply with your request for X-ray machinery having dimensions within the limits specified in your enquiry. The standard apparatus for diagnostic work produced by this Company is...

...It is possible, however, that a special machine could be produced, capable of fulfilling your requirements. Any order for such a special machine would necessarily involve a substantial increase in costs.

This is 'correspondence English' in its most deadening form. Our aim should surely be to create the impression of an efficient, friendly organization of human beings, in which there are technical specialists interested in finding solutions to the customer's problems. This calls for a clear, friendly, unaffected approach:

Our usual X-ray machines would be too big to fit into the space you have available; but we could certainly make a special machine...

The features of style that I have stressed already in this book apply as much in letter-writing as in any other writing. Directness and activeness should be the keynote:

Write: I enclose a sample of XY 123 for you to evaluate.
Not: A sample of XY 123 is enclosed for your evaluation.

Write: We are developing this method of spraying.
Not: This method of spraying is actively receiving our attention with a view to further development.

Write: We suggest you try to overcome this difficulty by...
Not: It is recommended that an attempt should be made to obviate this difficulty by...

In a serious conversation with a customer or contact on the telephone, you would not say: *It would be appreciated if you would favour us with your comments.* The comfortable rhythms and vocabulary of direct speech would produce something like *Please let us know what you think.* These words are easier for the receiver to assimilate; there is nothing inaccurate, slangy, or discourteous in them; they are therefore preferable in writing as well as in speech.

I am not for one moment advocating the use of loose, slangy, chatty diction. My attack is as much against the jarring effect of sudden lapses into racy 'mateyness', which offends by its presumption and imprecision, as against the blurring effect of over-formal phrasing or unnecessary jargon.

Try to think that your client or customer is sitting right in front of you; then write as far as possible in the words you would use in a full, direct and precise account of your message. In this way, you should avoid erecting an unnecessary barrier of 'correspondence words' between your customer and your message.

Correspondence words are words that somehow seem to take over from perfectly good everyday words as soon as people begin to write letters:

Use	Don't use
tell	acquaint/advise/inform
begin/start	initiate/commence
send	despatch
buy	purchase
soon	at an early date
because	due to the fact that
we can	we are in a position to undertake
about	concerning/regarding/with respect to/
	in connection with/in relation to/with regard to

These words and phrases are undesirable not because they cannot be understood, but because they build up an unnatural, stiff, unnecessarily formal atmosphere about letters.

Why do writers use this style? It seems that they believe that letter-writing calls for greater dignity or formality than discussion or conversation. This is almost true, but not quite. In writing, we certainly need to be **more explicit**. We have to put into words some elements of our message which in speech we could rely on passing by intonation, gesture or other signal. Also, we have to be sure that the form of words we use in writing will be entirely comprehensible to all our readers without ambiguity, and without further explanation. In contrast, in speech we can try out one form of statement, and if we see that it has not been comprehended, we can rephrase it immediately, expanding it and reorganizing it until our listeners understand.

Writing calls, then, for greater precision and greater fullness of detail: but it does **not** call for a more roundabout, more pompous, more long-winded style.

Personal or impersonal?

To keep letters as natural and direct as possible, use the first person, singular or plural, whenever it seems comfortable to do so. This does not lay you open to personal responsibility for the information you give; use of the first person will not make you more or less responsible. You write as the company's agent anyway; you use notepaper headed with the company's name and give your office in the company after your signature. If you sign a letter which makes a recommendation, you will not decrease your personal involvement by using the more awkward, evasive phrase It is recommended that.... Other avoiding phrases such as *the undersigned* or *this office* or *the present writer* are just as awkward:

> The writer is aware of the use you have made of A, B and C for various lubricating purposes over the past years, but we would advise you that X will fulfil all the functions hitherto performed by...

> Therefore, you are requested to communicate initially with the undersigned at this office.

Notice how the sentence structure of the first example is contorted to accommodate *the writer's* unnecessary attempt at self-effacement. To me, the unnatural construction *the writer is aware of the use you have made* is far more obtrusive than the simple *I know you have used* would have been. Similarly, in the second example, the artificial expression *communicate with the undersigned* jolts me into noticing the style used in the writing, when I ought to be concentrating simply on the message.

Probably the most natural way of writing is to use the first person plural, emphasizing that you are a member of a company of lively people. You can then vary your constructions and use the first person singular or the impersonal to gain variety or particular effect:

> We have three experimental constructions set up to test these procedures. If it would help you to come to discuss these with us, I should be pleased to see you any afternoon next week.

Courtesy

One further point we must bear in mind is ensuring that we give the right impression of courtesy. Naturally, in writing to customers, we must pay particular attention to being courteous. Avoid being too abrupt, commanding or supercilious:

> You did not send us a large enough sample to test fully. Please ensure that samples sent for test in future are a mininum of...

or sounding tactlessly critical:

> It was unreasonable to hope that such a pressure could be achieved in your existing steam-producing plant.

> You must have misunderstood our instructions about...

But this does not mean that you should go to the other extreme and become unnaturally deferential or obsequious:

> We would respectfully beg to inform you...

...accept our humble apologies for the unfortunate incident and trust that this will not disturb the confidence hitherto reposed in us should we be given a further opportunity to...

Avoid, too, the inflating effect of artificial gush and dramatic flourish:

It is with great pleasure that we have to inform you...

We would venture to suggest to you...

Excessive courtesy rings false and hollow sentiment is much worse than none at all.

Beginning and ending letters

Some of the most awkward writing in correspondence comes in the opening paragraphs:

Adverting to your esteemed order of 24th inst. concerning nuts and bolts...

Re your letter of yesterday's date, reference AB/CD/1234,...

Further to your telephone communication of yesterday...

Thank you for your letter of 24th January 1991, reference AB/CD/ 12345, regarding the position in respect of your order for aluminium alloy window frames.

We enclose herewith our catalogue re our telephone conversation of yesterday's date entitled 'Polyacrylate rubber'

An excuse that is often offered for such clumsy opening sentences is that it is essential to mention the date, the reference number, and the subject of the customer's letter as early as possible in the first sentence. This is not necessary in a well organized letter. The customer's reference number should be put at the top of your notepaper, against a printed sub-heading 'Your reference'. This reference will often include the date and an indication of the subject of the letter; but if you are still anxious to make the subject plain, perhaps because you write many letters to the same customer on various subjects, put a heading immediately after 'Dear.....'.

Dear Mr Brown,
<u>Aluminium Alloy Window Frames</u>

This leaves only the date of the customer's letter to be mentioned in the opening sentence, and this can usually be done unobtrusively:

Half the nuts and bolts you ordered on the 24th January are now ready and...

There are two possible solutions to the problem you raised in your letter of 24th January...

After our telephone conversation yesterday, I checked...

Above all, don't set your letter off to a bad start by opening with a grammatical error:

With reference to your letter of 4th January. We are sorry...

In reply to your letter of 7th December. There is still time...

These are incomplete sentences. If you open with *In reply to*, you must go on to state what you are doing or have done.

Begin in a crisp, businesslike way with your first point. Avoid the stereotyped, cumbersome opening:

Thank you for your letter of 28th June 1991, reference AM/Trg/14, and your enquiry regarding the possibility of our presenting Effective Communication courses in connection with your company at the Grand Hotel, Seaside. I am writing to inform you that negotiations with that hotel have now been brought to a successful conclusion and we are able...

Begin with the first point you wish to make, particularly reserving the opening *Thank you* for occasions on which you really wish to thank people for a particular service or favour:

Negotiations with the Grand Hotel, Seaside, have now been completed and the only matter still to be decided is...

Thank you for sending me an extra copy of the agreement on...

Would it be possible to change the date...? One of our staff is seriously ill, and...

Finally, you have to end your letter. Don't feel that you must finish with a flourish. If you use one of the customary formulae, your flourish will be only too plainly an empty gesture:

We trust that this information will be of use to you, and look forward to being of future service.

If you would like further information, please do not hesitate to contact us.

Assuring you of our best attention at all times.

We await the pleasure of your further instructions.

Yours in anticipation of an early reply.

We look forward to hearing from you at your convenience.

The point about these sentences is that they are now so over-worked that they are merely empty noises. If you really want a particularly prompt answer, ask for it directly and say why. Otherwise, avoid the coercing tone of the last two examples (and the probable snigger at the double meaning of the last). It is much better, when you have finished your message, simply to stop. Most people will not feel that this is unduly abrupt: they are more likely to get a sense of clear thought and crisp control.

When you are an enquirer

This discussion has been concerned with letters written in response to a previous contact – to answer an enquiry, to give information following a test, or to comment on a suggestion. These will probably be the main part of your day-to-day correspondence. But sometimes you will be a customer or enquirer yourself. The need for care and exactness is just as great. Make plain what you want to know. A vague question invites a vague answer. It would be difficult to give a direct and helpful answer to a question such as:

I am interested in switches for a new thermostatically controlled electric radiator. Please send me details of any appropriate products in your range.

CHAPTER SEVEN

Writing for international audiences

7.1 THE NEED TO ACCEPT RESTRICTIONS ON LANGUAGE

The theme of this book so far has been the need for technical writers to write accurately, directly, and sensitively – with *sensitively* meaning 'with due attention not only to the accuracy but also to the manageability, readability and acceptability of what they write'. I have stressed that this entails using language flexibly, selecting from the widest possible variety of vocabulary and structures just the right language to give the desired meaning, balance and tone. In particular, I have condemned the distorting effect produced by efforts to straitjacket technical ideas entirely within impersonal, passive, past-tense constructions.

But there are circumstances in which it *can* be beneficial to restrict the forms of English used in writing. In documents that are to be read mainly or largely by people for whom English is a foreign language, careful control of the range of language can help the readers considerably.

For convenience, in the rest of this chapter, I shall use the abbreviated terms 'native readers' and 'foreign-language readers' to save repetition of the lengthier descriptions 'readers who are native speakers of English' and 'readers for whom English is a foreign language'.

Native speakers of English must recognize that delicate distinctions of meaning or subtle shades of tone may not be recognized by foreign-language readers. Foreign-language readers may not understand that the phrase *a novel technique* is intended to convey more than that the technique is new: they may miss the intended implication that it is 'ingenious, rather clever and interesting' as well as new. Foreign-language readers may find it difficult to recognize that the statement *production is to all intents and purposes static, tending if anything to decline* is intended to mean 'production is falling slightly'. They can be forgiven for not knowing that *it could be said that* usually is a self-protective way of saying 'I think': the use of *could* certainly does not imply a particular condition. And they are almost certain to have difficulty in grasping the meaning of the sentences below – most native readers do, too:

The contribution of the series resistance of the electrolyte towards dissipation factor decreases with decrease in capacitance (for the same case size), therefore there are lower losses in capacitors with the higher rated voltages.

Since the degree of concavity is an inverse function of the degree of separation, and capillarity influences its height without altering its basic character, it follows that 'capillary manifestations in concave variants' is the correct description and prediction (generalization) of the character and behaviour of a sinal fluid level, and that these fluid levels are co-variant with the characteristics of space at the level of sinal pooling in all erect positions of the skull.

Authors may think they are showing a praiseworthy command of English if they make extensive use of statements intended to imply more than they say on the surface; or if they make extensive use of verbs such as *should, would, could, may* and *might* to impute varying degrees of certainty; or if they make extensive use of complicated sentence patterns. But extensive use of these features of language is not likely to impress foreign-language readers. For such readers, these features are more likely to cause difficulty and confusion. For effective **international** communication, especially for communication in commerce, industry or research, it is wise to restrict the range and the complexity of the English used.

Indeed, it is possible to restrict the English used in written documents so carefully that the documents can be 'read' by people who do not truly 'understand' English at all. These people do not 'read' English: they simply recognize patterns of familiar symbols, or marks and lines on paper. Yet, they can be taught to work efficiently from documents written in this restricted range of symbols, in this 'controlled' English (described in section 7.4). There are obvious commercial advantages in being able to distribute internationally just **one** controlled-English version of a piece of information, instead of having to publish many versions in different languages.

Of course, it is necessary to use a complete range of English to express complex or abstract ideas. Controlled English **can** cope with commercial and technical information like installation instructions, maintenance and repair instructions, operating procedures, and various types of descriptive writing. Controlled English can **not** cope with theoretical discussions, arguments about data, or with very abstract analyses. Consequently, there is much new technical information that manufacturers would like to distribute world-wide in just one version, but that controlled English cannot accommodate. For that information, translation is necessary. Nevertheless, even if we have to present complex information, we should keep the principles of control clearly in mind as we prepare to communicate with *international* audiences.

7.2 WRITING FOR 'EXPERT' READERS

In international communication in commerce, industry and research, there are three broad groups within the audience for information in English. For convenience, I shall call them the expert group, the student group, and the no-English group. (I should stress here that I am confining my discussion to international communication in commercial or professional activities. I am not discussing 'English for tourism' or the development of communication skills for general social and cultural activity, which would introduce other groups with other needs and motives.)

The **expert** group is probably the largest of the three. It consists of:

1. people who want to read international journals and books about science, technology, medicine, computing, commerce, legal matters, or other professional subjects in English;
2. people who are working with internationally available complex equipment, for which the support documentation is in English;
3. people who are involved in international commercial activities, in which the common language used is English.

These people read books or listen to lectures in English because they want to add to the expertise they have already in the subject-matter of the books, the operation of the equipment, or the activities of the international organizations with which they are connected. They may have learned their fundamental knowledge in their own language or in English. For many, the specialist vocabulary they have learned is the same in English and in their native languages. Those who are not so fortunate have to learn the equivalent specialist words in English and their own languages. This is hard work, but it is not usually confusing. There are usually reliable direct translations between the two languages. For expert readers like these people, it is not usually the special terminology in English that causes trouble. It is the 'ordinary' language in between.

Specialist terminology, and 'the ordinary language in between'

Let me stress the importance of 'the ordinary language in between' by inviting you to read through two examples from technical writing in French. In the first example, even if you know no French, and even if you know little chemistry, you will be able to recognize many of the specialist terms:

1.   *Décomposition du maléate de dimétindène à l'état solide*
1.1   *Enchantillons*

Afin d'éviter une décomposition trop rapide et incontrôlable du maléate de dimétindène, ce dernier est mélangé avec de la cellulose microcristalline dans les proportions (1 + 4). Des èchantillons de dimension différente sont ensuite préparés, en fonction de chaque test de stabilité.

1.2   *Conditions de conservation*
1.2.1   *Influence de la température, de l'humidité relative (hr) et de la lumière*

Des échantillons de 500 mg sont distribués dans des piluliers de 18 ml en verre blanc et conservés à l'abri de la lumière à 85°C et 70% hr, 55°C et 60% hr, 40°C et 60% hr, ainsi qu'à température et humidité relative ambiantes, à la lumiére du jour. Pour l'analyse, un échantillon est repris par 3 ml de méthanol, agité et centrifugé.

In the following example, not so many of the terms are immediately recognizable; but once you have learned from your dictionary that *donnees* means 'data', *ordinateur* means 'computer', *traitées* means 'processed' (traitement de texte = word-processing), *le bottin* means 'telephone directory' (or 'phonebook'), and *clavier* means 'keyboard', you can begin to make some sense of the text:

DONNÉE
Ce sont simplement les informations fournies à l'ordinateur (données d'entrée) ou encore les données reçues de l'ordinateur (données de sortie ou traitées).

Ainsi si je prends au hasard un certain nombre de noms et d'adresses dans le bottin et que je les rentre en mémoire de l'ordinateur (en frappant sur le clavier), j'alimente l'ordinateur en données. Si je lui demande ensuite de me trier ces données par ordre alphabétique (ce qu'il peut faire sans grande difficulté) j'obtiens des données traitées.

Although consulting a dictionary is a tedious task, gradually we assemble reliable definitions for individual words: *échantillons* means 'samples'; *mélangé* means 'mixed'; *pilulier* means 'pill-maker'; *alimente* means 'feed', and so on.

We are alerted by the dictionary to possible pitfalls. In the first example, *verre blanc* does not mean 'white glass' but 'plain glass'. In the second example, *trier* does not mean 'try' but 'sort'. These are what linguists call *faux amis*, 'false friends', words that look as if they will mean the same in two languages, but actually do not.

(*Actually* is itself a 'false friend'. A French reader who did not have a good command of English might be tempted to think that *actually* means the same as the French *actuellement*, and would interpret the last sentence of the previous paragraph as 'currently (or at the moment) they do not'.)

But after we have obtained equivalents in English for all the main French words, our major problems start. We have to interpret 'the ordinary language in between'. In the heading of the first example, we have to deduce the precise meaning of the preposition *à l'*. The dictionary tells us it can mean 'to, into, at, on, during, till, until, in, by, per, from, out of...' (there are 17 entries, occupying more than a page of small print, for 'à' in Robert & Collins *Dictionnaire Français – Anglais, Anglais-Français*). So, does *à l'état solide* mean 'to the solid state' or 'from the solid state'?

The dictionary tells us that *ainsi* means 'in this way' or 'thus'; and *ainsi que* means 'as' or 'just as'. But when we try to understand the expression *ainsi qu'à température et humidité relative ambiantes*, the dictionary meanings do not seem to fit together to make good sense, especially when we try to relate the *ainsi...* group to the rest of the sentence.

As you can see, it is not the main technical terms that are our main source of difficulty; our main sources of difficulty are the structural words and the structural signals (like word endings) that are vital indicators of how the main words are supposed to be related to one another.

When we are writing technical texts for distribution throughout the world, it is easy for us to overlook the distinction between expertise in a technical subject and expertise in English. It is salutary for us to be reminded of the difficulties **we** face when we are confronted by a text in a language other than English.

Certainly, we can usually assume that an expert audience will have little difficulty in understanding the established core of specialist terminology in our subjects. (I am well aware that it is not easy to describe 'the established core' exactly, and I agree that writers should err on the side of explaining slightly too much rather than slightly too little.) But we must remember that, even for expert audiences, the most frequent causes of difficulty are not single specialist terms but complexities of sentence structure, complex verb forms and apparently common English words and phrases that native speakers of English use almost without thought.

Structural complexity

Consider, for example, the following paragraph (the original was a continuous text; I have numbered the sentences to make discussion easier):

Sentence 1   In such a case, the program would be run once and provide one set of results.

Sentence 2   In fact, it is the simplest form of program possible except for one which does not have a repetitive loop.

Sentence 3   To make the program more useful still, we should have to put in an input section in which we could read in different sets of values of A and B.

Sentence 4   This would necessitate the use of another loop.

Sentence 5   We might then decide to add the facility of changing the range of X and the X increment.

Sentence 6   This would introduce another input step and another loop.

Sentence 7   The possibilities are endless.

People who read this text must know a great deal about computing already, otherwise they would not be reading it. For them, specialist words like *program, repetitive loop, input,* and *increment* are not likely to cause difficulty. Much more difficulty is likely to come from the mixture of positives and negatives in the second sentence *simplest... except...not.* More difficulty is likely to come in the third sentence from the phrase *more useful still* and from the shades of meaning of *should* and *could*. More difficulty is likely to come from the optional use of *might* in sentence 5. And some will find it difficult to comprehend the final idiom *the possibilities are endless.*

Again, in the following extract, the main source of difficulty is not likely to be the specialist terminology:

At full loadings the output voltage and current are the combination of two components in quadrature, one from the secondary winding of the transformer T.2 and the other from the secondary winding of the transformer T.1. With this circuit, if for a given value of load resistance R, the reactance of the transformer T.2 is minus R and that of T.1 plus R, then the load current of T.1 is the reactance current of T.2 and the load current of T.2 the reactance current of T.1, thus the former leads the supply voltage by 45° and the latter lags by 45° and the two load components are thereby 90° displaced. The output has a two phase character, whilst the input current is in phase with the supply voltage.

In this example, the greatest source of difficulty, I suggest, is the unnecessary complexity of the sentence structure.

The following examples again show difficulties arising not from the specialist vocabulary but from thoughtless writing. Here, the examples show writers not recognizing the difficulties caused for foreign-language readers by informal phrases like *will remain tied up*:

If all aircraft are stored with tanks two-thirds full, vast quantities of fuel **will remain tied up** for long periods.

The X Doppler system **is thrown in** mainly to...

If the system **does not come up with** your requirements...

The suggested delivery dates **will be tight**.

...but since the X system is **quite** new...
  (meaning **entirely**)

You should be able to identify the programs that are **lifted** directly from the Pascal system.

At the end of the test, the connection between the two modems **is dropped**

In this section so far, I have mentioned three ways in which native-English writers produce text that presents difficulties for foreign-language readers:

- by creating complex sentence structures;
- by creating complex verb forms, especially forms using auxiliary verbs like *should, could* or *might*;
- by using informal, 'colloquial' language.

Elsewhere in the text, I have stressed the ways in which various other features of style create difficulties for foreign-language readers. To ensure that all those ways are gathered together in this chapter, too, here are some more examples (with references to the pages where the features of style have been raised previously):

Use of 'fashionable' words that have several meanings
(pages 25 to 27)

...because of the enhanced risk associated with the supply of X.

This enhances the revenue potential of aircraft A by several...
...the turn-around time of aircraft B is enhanced.

Is an *enhanced* risk an increased risk or an improved (and therefore reduced) risk? To *enhance* the revenue potential of an aircraft is to increase that potential; to *enhance* the turn-around time of an aircraft is to improve (and therefore reduce) the turn-around time. These examples cause native readers to stop and think; for foreign-language readers, even expert translators, they are considerable obstacles.

A related cause of trouble is inconsistent use of terms. I have a manual that switches confusingly between *fixed disk* and *hard disc* to describe a single unit.

Another form of inconsistency is shown by this example:

Entering the command, ABC7, in the XYZ domain causes you to leave the XYZ domain and **enter** the ABC domain.

In computing, *enter* is used to express at least four meanings:

- to type;
- to press a key (Enter or Return) to transmit to a file the data you have just typed;
- **both** to type in **and** to press a key to transmit data to a file;
- to move into (as in 'enter the ABC domain').

How should translators translate *enter*? By using just one word in their own languages (and thereby ignoring/compounding the confusion), or by attempting to identify the exact meaning of each occurrence of *enter*, and supplying a different, explicit word for each occurrence?

Be consistent in your choice of terms. Choose one word for the activity of typing, and use it consistently to express that activity. Choose one meaning for the term *enter*, and use *enter* consistently to express that meaning.

Pre-modification
(pages 32 and 33)

Here is an example of a **complete** sentence:

Correcting transfer voucher procedures are in Chapter X.

That example of a **complete** sentence can be condemned on several grounds: abnormal word-order, excessive pre-modification, pre-modification with nouns, pre-modification instead of use of a more comfortable prepositional construction. I had to re-read the sentence several times before I worked out that it meant:

Procedures for correcting transfer vouchers are in Chapter X.

I suspect that foreign-language readers would have to think very hard about how to translate that awkward wording.

Mis-related phrases and clauses
(pages 75 to 77)

When emptying the system, filtrate from tank X can be fed to the top of reactor Y.

By pressing ES, a list is obtained.

To monitor the conversation, the loudspeaker must be operated.

Foreign-language readers, decoding in accordance with the rules of English grammar that they have been taught, are more likely than native readers to find those expressions confusing, precisely because they have been taught the formal rules of English grammar. Often, they are able to see and explain a mis-relation which native readers, many of whom have never been taught the rules of English grammar, are dimly aware of but cannot explain.

Omission of punctuation
(pages 78 to 83)

Label your volume when prompted with an easy-to-remember name of up to 11 characters.
(when prompted, with)

Selection is controlled by three colour enable lines.
(colour-enable)

...are responsible for customer collections.
(collections of customers?
collections by customers?
collections from customers?
customers' collections?)

Expressions such as the first two of these examples cause a slowing-down of a translator's activity. Expressions such as the last cause a complete stoppage.

Tone
(pages 84 to 91)

You are strongly recommended to attend an ABC system generation course, if possible; it will probably save you a lot of time (and possibly heartache).

To recognize the difficulty created by this attempt at humour, it is instructive to put yourself in the position of a foreign-language reader who has to look up *heartache* in a dictionary.

If you were a French reader, you might look up *heartache* in your Robert & Collins *Dictionnaire Français-Anglais, Anglais-Français*. You would find that the equivalent words in French are given as *chagrin* and *douleur*. So you would interpret *heartache* as having the same meaning as is attached to those words in your French context. But what are the meanings attached to those words in a French context? At the other end of the French-English, English-French dictionary, *chagrin* is defined as *grief, sorrow, distress*, and *douleur* as *pain, grief, distress*.

Did the writer seriously wish to suggest that a course on system generation would cause grief, sorrow, distress, or pain? Almost certainly not. (If he did, his colleagues responsible for the course would have good cause for resenting his published comment!) Most native readers are able to detect his intended nuance of jocularity; but foreign-language readers need a very high competence in English to pick up nuances of that sort. And even if they could, would they feel that it was appropriate to have jocular asides in a technical manual? Attitudes to humour vary widely in different cultures.

Differences between British English and American English

Writers in English must even take care about the words and expressions they use in communication between England and the USA. The Englishman who is visiting a laboratory in the USA and is asked 'What time do you have?' will find his reply 'Not much, I'm afraid' greeted with some confusion. His answer should be based on a quick reading of his watch – 'Ten minutes past three' or 'Ten after three'.

It is important to remember that there is not just one English language: there are (at least) British English, US English, Canadian English, Australian English, and South African English. These languages are marked principally by similarities; but there are many differences, too. If we want high reliability in the transmission of information, and if we wish to avoid causing offence by creating inept tone, we must ensure either that all our texts are written in the language of the target audience, or that the target audience is fully alerted to the points of difference in the code that is being used.

It is interesting to recognize that it is not the big and obvious differences that cause difficulties in transatlantic communication. For example, pronunciation and intonation are obviously very different in British English and US English; but after a tuning-in period, we do not often find that pronunciation and intonation differences cause complete breakdown in communication.

There are some differences in labels that must be learned: for example:

US English	British English
faucet	tap
ground	earth
thumb tacks	drawing pins

antenna	aerial
hood	bonnet
in back of	behind
initiate your main lamps	switch on your headlights

but we soon learn these, because if we do not, communication stops.

Much more dangerous are apparently common words or idioms with which we assume everyone on the receiving end is familiar: communication goes on, with participants unaware that there are differences of interpretation between transmitter and receiver:

> US use **alternate** equipments
> (British: alternative)

(To a British reader, the usual meaning carried by *alternate* is 'first one, then the other, then back to the first'.)

> US Drive on the **pavement**, walk on the **sidewalks**.
> (British: Drive on the **road**, walk on the **pavement**)
> US The **regular** operator
> (British: The usual/ordinary operator)
> US If *target* is a **regular** file, its contents are destroyed.
> (British: is a normal file)

For British readers, the primary meaning carried by *regular* is 'recurring at fixed intervals or in a repeated, consistent way; in accordance with a habit, order, rule, or custom'. Accordingly, British readers would interpret *regular* food in the next example to mean 'food taken at consistent and evenly spaced intervals':

> In cases where the patient cannot eat enough, supplements of X in liquid form, taken with regular food, will provide the calories and protein required...

The American writer did not intend that meaning. Her intention was to speak of 'normal, usual' food.

Of course, many readers become familiar with differences of use on the two sides of the Atlantic; but if readers have not had much exposure to the 'other' form of English, there is considerable danger of native readers, let alone foreign-language readers, missing a difference of meaning.

We need to be on the look-out not only for differences in the meanings of single words but also in the implications of larger groups. To a British reader, the following extract looks like a typical piece of passive, impersonal, technical writing, meaning 'When (while) formatting is taking place, you will see the prompt':

When formatting is done, you will see the > prompt.

In fact, the American writer meant:

When formatting has been completed, you will see the > prompt.

In British English, the presence of *Have* at the start of a sentence always signals the beginning of a question. It is therefore a surprise for a British reader to read:

Have your assistants put the pipe squarely into the flange...

Have the recycle systems regenerated. Then,...

Unless he or she is familiar with this common American construction meaning 'Ask your assistants...', 'Ask the operator to...', 'Ensure that...', 'Get the recycle systems regenerated', the British reader may well wonder if a question mark has been omitted by error. Worse, to a British reader, there is an air of authoritarian compulsion about:

Have them inspect the sample first.

The American writer's meaning was:

First, show them the sample.

The sentence was intended as the neutral statement of the beginning of a procedure. No authoritarian tone was intended.

There can be difficulties over the use of abbreviations, too: for example, over the American use of # to mean number, over the European use of the comma as a decimal point, and over the American way of indicating dates (3/12/91 = 12th March 1991).

So, for expert readers overseas, even in the USA, writers should constantly keep in mind the nature and range of English likely to be at the disposal of the audience. It will almost certainly be desirable to limit the language forms used and to make explicit many nuances of meaning and tone that could be expressed to native readers by subtleties of structure or vocabulary.

7.3 WRITING FOR STUDENTS

In the student group within the international audience, I include all people who are studying English itself or who are using English as the medium of study of commercial, scientific or technical subjects. These people have not yet learned the fundamentals of their subjects, even in their own languages. They may have to build up in English the whole conceptual framework underlying each special term. When we plan to write or speak to such readers, we must make great efforts to give clear definitions and explanations.

This does not require different skills from those we need for writing for expert readers. The task differs more in degree of difficulty than in kind. The main difficulty is still to control carefully the complexity of the structures created. We must make even greater efforts to control the language used and the rate at which information is unloaded. The greatest difficulties for student readers are again caused by difficult phrases, complex sentences and shades of meaning expressed by *would*, *should*, *may* and *might*.

7.4 WRITING FOR READERS WHO DO NOT UNDERSTAND ENGLISH: USING 'CONTROLLED' ENGLISH

When we come to present information to the third category in our audience, to the no-English group (the group who do not know how to speak or write **any** English), we have to exercise the greatest restriction and control. Yet, and this may seem strange, provided we know that these readers have been trained to recognize the words (symbols) we use, we can be more confident about communicating clearly with these readers than we can with either of the other groups. When we communicate with experts or students, we can never be absolutely sure that we have judged accurately their ability to understand us. In contrast, if we know our readers have been taught to recognize the symbols we use, we can be confident of clear communication.

Several 'controlled' versions of English have been developed, by large international companies like the Caterpillar Tractor Company, Digital Equipment Corporation, L.M. Ericsson, IBM, Eastman Kodak, and Rank Xerox. All are based on an original idea worked out by the Caterpillar Tractor Company of Peoria, Illinois.

The Caterpillar Tractor Company distributes tractors and heavy earth-moving equipment throughout the world. It supports its machinery with maintenance and repair documents, and because this has to be done in many countries, the company used to provide the documents in many languages. But duplication of documents in many languages is expensive, so the company looked for alternative ways of presenting its information.

Caterpillar research workers decided that it should be possible to use a single, internationally understood set of symbols to convey much of the information they had to transmit. And having decided this, they explored the possibility of using English words as this set of symbols. They produced a list of 784 words as a central core of symbols, plus a list (with many illustrations) of names of parts of Caterpillar equipment.

The research workers found that they could express all their service and maintenance information using this vocabulary alone in a carefully controlled range of simple structures. They found, too, that in 30–60 hours they could train operators, who previously knew no English, to recognize the meaning of the documents written in this way. Though the operators did not 'understand' English, they could work efficiently on the basis of the information drawn from the controlled-English documents.

The Caterpillar Tractor Company named its restricted version of English 'Caterpillar Fundamental English'. It is marketed now as ILSAM – International Language for Service and Maintenance – by M. and E. White Consultants (world-wide agents) and as BASIC 800 by Smart Communications Inc. (agents within the USA only).

The principles for producing a restricted or controlled version of English such as ILSAM or BASIC 800 are easy to grasp. The variety of words used must be strictly limited, and each word must have one meaning only. For example: *right* is the opposite of *left*, *correct* is the opposite of *wrong*. *Drop* is a noun meaning 'a quantity of fluid that falls in one spherical mass'; it is not used as a verb meaning 'to fall' or 'to release'; and it is not used as a noun, as in *a drop in pressure* – that is a *decrease*. The word *over* is restricted to the single meaning 'above', as in *placed over the valve*; it is not used to mean 'more than' as in *produces over 10 watts*, or 'during' as in *over the three days*, or 'finished' as in *that the emergency was over*.

Synonyms are avoided: from several words that have approximately equal meaning, one only is chosen for use. For example: *below, under, beneath* and *underneath* all have similar meaning. ILSAM uses only *under*. Also, wherever possible, the word with the widest international recognition is chosen for use: *assistance* is used, not *help*.

The number of verbs is kept to a minimum. This is achieved by use of verb-noun combinations as much as possible: *make an alignment* is preferred to *align*. This has the virtue of reducing the number of verbs to be learned and of reducing the number of irregular verb forms to be used.

Statements are made as short and positive as possible.

Not: The control unit, duplicated for safety, has a low resistance.
But: The control unit has a low resistance. There are two units for safety.

Repetition replaces reference back, and explanations are carefully sequenced in steps. Tense, voice and mood of verbs are carefully restricted. Wherever possible, sequences of statements, plus words like *before, after, last, first, then,* are used instead of past and future verb forms.

Writers can use all the four main types of sentence structure in English:

1. statements, descriptions or explanations;
2. instructions or commands;

3. combinations of conditions with either descriptions or instructions;
4. questions.

Statements can have qualifying words added, but not too many. If necessary, adjectives can precede both subject and object. The first sentence below has no adjectives: the remaining sentences add more:

The washers prevent leaks.
The network uses reed switches.
The control system has two complete stages.
Twelve GV-1 groups are normal in group selectors.

Heavily qualified statements are avoided.

In instructions or commands, each sentence contains just one instruction:

Not: After stopping the program, load the data into the buffer store.
But: Stop the program. Load the data into the buffer store.

Not: Using program 6, send X to Y.
But: Use program 6. Send X to Y.

Clauses giving explanations, conditions and indications of time can be combined with descriptive statements or instructions:

Reason clause: Because you must make the loop first, the linkage is important.
Time clause: When the work is complete, put the test line into the cable duct.
Condition and
  instruction: If the error is larger, increase input at X.
Condition and
  statement: If the interval is less than three seconds, the sequence is wrong.

But these combinations must remain manageable: one qualifying clause per statement is a good rule.

Questions are constructed as simply and directly as possible:

Can the door open freely?

Of course, non-English-speaking operators cannot just pick up and read documents in ILSAM or any other controlled version of English, however carefully the documents are written. A training programme is necessary. In the training programme, a bilingual instructor helps the operators to recognize and understand the significance of the limited vocabulary and the limited range of patterns. The instructor gives definitions and explanations in the native language. At no point is the learner required to speak or write English. Caterpillar have found that a course lasting 30–60 hours is normally enough to enable 'readers' to work competently from ILSAM documents.

An important feature of this type of controlled English is that it is not distorted or artificial. The vocabulary and structures it uses are selected from those used daily by native English speakers. Documents in this language are therefore entirely acceptable to native English readers. Indeed, when the Caterpillar Company published the first service literature in its restricted language, the difference was not detected by native English readers!

Here are some examples of texts re-written in 'Ericsson English' – a controlled version of English produced by the University of Wales Institute of Science and Technology for the L.M. Ericsson Telephone Company, Stockholm [13, 14, 15]. The verbs are in italic. The original texts use relatively complicated verb forms such as *have been positioned* or *being untwisted*, and the auxiliary verbs *may* and *should*. Passive contructions and auxiliary verb forms are major sources of difficulty for readers who are not proficient in English, and they are not necessary in simple instructions and explanations. Note, too, how it is possible to have different re-written versions: use of controlled English does not mean that writers have no choice in the language they use.

1. *Original text*
   Stripping of cables for normal magazines
   When the cable *has been positioned* on the cable shelf for normal magazines, *use* measuring rib 1007478 to mark where it *is to be* stripped. Stripping *is to take* place immediately before connection work as the twisted groups *run* the risk of *being untwisted* if the cable *is stripped* in advance.

   One re-written version
   Stripping of cables for normal magazines
   Put the cable into the correct position on the cable shelf. *Use* measuring rib 1007478 *to make* a mark on the cable. The mark *shows* where to *start* stripping. *Strip* cables a short time before connecting, *to prevent* untwisting of cable elements by accident.

   Another re-written version
   Stripping of cables for normal magazines
   Make sure that the cable *is* in the correct position on the cable shelf for normal magazines. *Use* the measuring rib 1007478 *to make* a mark on the cable, so that the mark *shows* where *to strip* the cable.

Do not strip the cable until you *are* ready to start connecting. If you *strip* the cable before you *are* ready, you *can untwist* the twisted groups by accident.

2. *Original text*

In order *to be able to keep* the time schedule, it *is* important that the preparations *are* well *made* and carefully *thought* over (items 3.2 and 3.3), otherwise a waste of time *will* easily *ensue*. As *is* evident from the suggestion for time schedule below, several different tests *may be* in progress simultaneously. Just how long a time each test *will take is* dependent on local circumstances, but no test, with the exception of certain APZ tests *should take* more than 8 hours unless interferences *occur*.

One re-written version

We *make* careful preparations according to Sections 3.2 and 3.3, so that we *use* the period of the operation test in the best possible way. Figure X *shows* a plan of the length and sequence of the tests. The length of each test *depends* on local conditions. Several tests can operate at the same time. If we *do not find* faults, most tests *operate* for less than 8 hours. Some APZ tests *operate* for more than 8 hours.

Another re-written version

Important! *Be* careful when you *make* the preparations for the operation test (Sections 3.2 and 3.3). If you *are not* careful, you *will not use* all the available time successfully. Figure X *shows* that several different tests *can happen* together. The length of each tests *depends* on local conditions. If faults *do not happen*, the maximum time necessary for a test *is* normally 8 hours. Some APZ tests *need* more time.

Even if a company feels that it cannot use a single controlled-English version of its document(s) in all countries of the world – that is, that some translations are still wanted for diplomatic or technical reasons – the creation of controlled-English text(s) still has value. Since a controlled-English text has been written very clearly, with each word confined to just one meaning, it is ideal source material for translation. The translator can rely on the definitions of words, and finds the simple sentence-structure relatively easy to convert into comparably simple structures in the target language. Indeed, a reasonably accurate 'base' translation can be produced easily by computer, leaving human translators to add the expert touches – to add balance and polish by making the adjustments of word order and style required by the target language.

Principles similar to those used in creating ILSAM have been used to create controlled versions of English for international oral communication. For example, there are standard patterns of English for international aviation, and a Standard Marine Navigation Vocabulary has been produced by IMCO (Inter-governmental Maritime Consultative Organisation) [26].

The aim of the IMCO Vocabulary is to encourage all navigators to communicate effectively with one another by making them use an agreed range of words and phrases with agreed meanings. The Vocabulary suggests words and phrases for use in a variety of navigational activities. The words and phrases are arranged in groups under 21 activities, such as anchoring, fairway navigation, pilotage, radar transmission and fishing. There is a glossary explaining difficult terms, and an introduction describing how the Vocabulary should be used.

The design and use of controlled versions of English primarily (though not exclusively) for oral communication raise problems related to speaking and listening that do not concern us here. However, readers may be interested to explore the potential of controlled versions of English for international communication of commercial, scientific, and technical information in both speaking and writing. There is considerable scope for extension of the principles already used in technical documentation and in aviation and maritime activity.

7.5 ICONS

Perhaps, since use of words causes so much difficulty in international communication, we should abandon words wherever possible, and use icons instead.

It is impossible to make anything but very simple statements by stringing icons (images) together. Certainly, we shall not in the near future be able to use iconography in its present state of development to express complex descriptions, explanations, and instructions in our manuals or on-screen texts. But icons can be used as graphical symbols for objects, actions and events, and they can be used effectively in manuals and on screens to help readers with quick recognition and identification of information. For example:

- to help readers identify recurring categories of information (for example, recurring blocks of command-descriptions or operating instructions in chapters of a manual);
- to help readers identify quickly the content of files or other optional items listed in an on-screen menu;
- to help readers identify quickly the activities that will be produced by various commands to a computer;
- to help readers understand an event that has happened (for example, the linking of two computers).

Preferably, icons should be instantly comprehensible without explanation. It is acceptable to expect readers to be prepared to learn the significance of new icons, just as new words have to be learned when we begin to read about a new subject. But to be successful, icons must have qualities that make them easy to learn and remember. What these qualities are is still the subject of debate. As I write, the International Standards Organization is engaged in lengthy discussion about how to produce icons that will be recognizable internationally.

One problem is that 'familiar' objects do not have the same form in all countries and cultures. For example, which of the following icons is closer to your idea of a *waste bin*?

Which of these icons is closest to your idea of a *printer*?

I have no doubt that we shall gradually have to include more and more icons in our presentations of information, especially in our on-screen presentations. In general, to make accurate, clear statements with icons, we shall have to think along much the same lines as we use in planning accurate, clear writing. But there is an obvious visual element in 'good style' for communicating with icons, and discussion of that element would take us beyond the bounds of this book.

CHAPTER EIGHT

On avoiding ambiguity

I am often asked for advice on how to avoid ambiguity in technical writing. One answer I can offer is the use of carefully controlled language, limited in vocabulary and structures, as described in section 7.4. But such a form of 'controlled' English will not meet all our needs in writing technical texts. To communicate the shades of meaning and nuances of tone in subtle and complex description and argument, we need to use the full range of the resources of English. In such circumstances, I have to give a different answer to the question about avoiding ambiguity. The answer is: we can't.

As readers or listeners we can, if we wish, wilfully misconstrue almost any statement, especially one that contains general rather than special vocabulary. In reading an article about new designs for trains and stations, we may smile momentarily at the double meaning of the statement *The buffer will have an inflated head*. But it is really our own fault if we are distracted from the intended information about a mechanical device by an image of a human being with a peculiar physical shape.

Normally, we do not decode messages by looking for the **least** likely meanings that can be attached to the signals we receive. We do not receive and interpret the signals in isolation: we take into account the whole context in which they are being used. We consider the general linguistic surroundings of the particular signals; we consider the general theme that is being discussed, or the total object that is being described; we consider the writer's personality, education, occupation and experience, and the atmosphere pervading the communication situation as a whole. All these things are taken into account as we consider possible interpretations. This means that sentences including the following groups of words, which some people would condemn as ambiguous, would normally give no trouble:

> ...acquires a pitch that is in proportion to the distance the bat is away from the reflecting object.

> These stages consist basically of staircase generators.

The meanings we draw from *pitch* and *bat* vary according to whether they are included in a discussion of ultrasonic guidance systems (as here) or a sports report; and indeed their sporting meanings depend on the game being discussed and on whether the context is British or American. No reader is likely to be seriously confused. Likewise, no reader of the electronics journal from which the second extract was taken would deduce that the stages consist of machines busily producing staircases!

So in this chapter I want to stress the impossibility of avoiding ambiguity **entirely** in the use of words. But let me make clear what type of ambiguity I am referring to when I make that statement: I am referring to the ambiguity that arises when we can make any of several possible responses to correctly formed and correctly presented code signals. My purpose here is not to excuse the types of ambiguity that arise from 'common errors' in sentence construction, syntax and punctuation – that is, from ill-formed and ineptly presented code signals. For example, I am not excusing ambiguity that arises from misused participial constructions.

> The problem of contamination by bacteria and yeast was a major difficulty and after running for about two weeks the mammalian cells were found to be non-viable.

I am not condoning ambiguity that arises from careless use of pronouns:

> Samples were sent to Technical Service Department for evaluation and stored for shelf stability. They have reported good adhesion and corrosion resistance and suitable shelf stability.

And I am not accepting ambiguity that arises from slovenly punctuation:

> ...counts of the bacterial load in the air of piggeries. He observed that bacteria carrying dust particles decreased in concentration...

These are not difficulties of semantics: they are weaknesses in thinking, or defects of self-discipline in the writers. All good texts on efficient writing give copious advice on how to overcome them.

Nor, when I acknowledge the unavoidability of ambiguity, am I suggesting that we must condone woolly communication, which is unclear because a writer is tautologous, irrelevant or deliberately evasive, or because he or she fails to be specific in giving instructions:

> ...should take steps to overcome...
> (what steps?)

Disambiguation

My concern is to acknowledge the extent to which the communication process relies on the audience's ability to disambiguate what the writer is trying to say. It is unreasonable to expect complete freedom from ambiguity in the words we use in speech and writing. Constantly, we have to leave the receivers to decide which of a number of possible meanings are to be attached to the words according to their contexts. And even, on some occasions, when it would be possible to eliminate all chance of ambiguity, it is not desirable to do so, because it would involve a circumlocution that would unnecessarily lengthen (and even complicate) the receiver's task of interpretation.

For example, in technical documents it is usually quite acceptable to use terms such as *pitch, bat, gate, channel*, a *homing* device, or an *elastic* beam without elaborate explanation to ensure that misinterpretation is impossible. The essential skill in effective communication is judging what will be the familiarity of the receiver with the subject-matter and its customary vocabulary. The judgement that has to be made is not 'Could this word be wilfully misinterpretated?' but 'Is this word **likely** to be misinterpreted?'

Let me stress the importance of making this judgement. The language we use must be carefully and continuously adjusted to the capacities of the audience. My point is that often the audience is well able to select from possible meanings of the words used; the audience expects to have to **select** as a normal part of the information-receiving process, and does so without conscious effort. It is therefore unreasonable to give the impression that the use of words that are open to more than one interpretation is **always** unacceptable.

Unfortunately, some text books on efficient writing give this impression, and it has been a frequent source of unease to students I have taught. For example:

'Good diction means the absence of ambiguity, obscurity and misunderstanding' [27].
'To produce a sentence capable of one meaning only every time you write (or speak) a sentence must be your aim as a communicator' [28].
'Ambiguity is a deadly sin in writing or public speaking' [29].
'Language must be entirely clear to its readers. All distortions and ambiguities are abhorrent' [30].
'The scientific writer has an obligation to use words as accurately as he does numbers and symbols' [31].

I suspect that none of these text-book writers really believes that it is possible to avoid ambiguity completely; much that they write elsewhere in their books acknowledges the difficulty of being absolutely explicit; but their Olympian tone makes the learner-writer uneasy. Such advice is, I am sure, responsible for many of the ponderous linguistic contortions that inexpert writers indulge in as they attempt to be impregnably exact, complete, and (ironically) clear.

We all know from experience how difficult it is to be sure that our readers will distinguish between *discrete* and *discreet*, between *sine* and *sign*; we know that it is difficult to be sure that all our readers understand what is meant by a *dead* valve, or by *chatter* in machines; we know that the scientifically acceptable word *aperture* is no easier to define than the common-or-garden non-scientific word *hole*. We know, too, that to use the string of words *four plus four times three plus two squared* cannot be as clear as to use algebraic notation of numbers and symbols. So we must despair of being able to keep those text-book commandments.

The advice we give should surely be less dogmatic. We should acknowledge that it is difficult to express meaning exactly; we should point out that there are limits to the writer's responsibility for removing ambiguity; and we should suggest that the best way of minimizing ambiguity is to focus attention firmly on the needs and capacities of our readers.

Denotation and connotation

Our difficulties in exact communication of our ideas arise partly because words have both denotative and connotative meanings. Louis Salomon defines these terms in the following way:

> What we shall call the denotation of a word is the sum total of its referents: for example, *chair* denotes every single chair that has ever existed, or ever will exist, in the world of sensory experience. The connotation of a word is sub-divided into two parts: the defining qualities of the category or class it names, and the emotive or affective responses it arouses in the minds of its users. In popular parlance 'denotation' is often thought of as the real or proper meaning of a word, and 'connotation' as the mere accretion of illogical, even capricious feelings that help to mask or colour this meaning – hence, something that a well organized linguistic society would probably legislate out of existence. Note that we are making no such invidious distinction. **All** parts of the meaning of a word are there only because the users of the word impute them; denotation and connotation (both kinds of connotation) are equally respectable, equally important. [32]

This distinction between denotation and connotation is clear and important. But it is not simple. To complicate matters, a single word can have varying denotations in different contexts: for example, *scale* has different denotations in graphical representation, in zoology and in a grocer's shop; *wave* has different denotations in oceanography, in electronics and in a hair-dresser's shop; *pavement* has different denotations in England and America.

Some words have a definition but no concrete denotation. We have no concrete referent for *infinity, polarity, brittleness* or *parameter*. The referent for each of these is a mental construct, an abstraction or a perception of a relationship; scope for varying interpretations of these terms is substantial. Many of the terms we use in mathematics are understood entirely by their defining or informative connotations; and some terms have different informative connotations or definitions in different contexts: for example, *plasma* in physics and biology, *base* in chemistry and architecture, and *stress* in engineering and psychology.

Above all, it is the affective or associative connotations that give rise to ambiguity – to what William Empson called 'room for alternative reactions to the same piece of language'. [33] Consider the words *adolescent, progressive, latent, humid, consistency, significance* and *efficiency*. Can we be confident that we all have identical associations for those words?

Usually, in technical writing we want to minimize the possible variety of connotations, because we want to communicate about positive, concrete things – chemical substances, physical materials, units of machinery, instruments, reactions, states or conditions that must be carefully delimited. Scientists try to do this by using special vocabulary. W. E. Flood, in a study of the structure and meanings of scientific words suggests that:

> The scientist needs suitable names by which to identify the various abstractions with which he deals – processes, states, qualities, relationships, and so on. Thus, after Faraday had investigated the passage of electric currents through different solutions and noticed the resulting liberation of chemical substances, the term *electrolysis* was invented. This one word was a kind of shorthand symbol for the process; it 'pinned down' the process and conveniently embraced its many aspects. From then on it was possible to think about the process and to talk about it to others. [34]

But it is dangerous to follow Flood's simple theory and to accept that specially created words can subsequently be used with absolute confidence. He contends that:

> A scientist avoids the ordinary words of the language; he prefers his own words. These words can then be rigorously defined and given the necessary precision of meaning. [34]

It would be interesting to ask every reader to write a 'rigorous' definition of a selection of scientific and technical terms such as *calibration, hydraulic, lithography, matrix, parameter* and *transistor* to see with what agreement we 'pinned down' the processes, objects or states involved.

Our faith in the possibility of defining scientific terms rigorously must surely be shaken by Waismann's demonstration that we can never remove all possible doubt about the meaning of words, even of terms denoting physical objects:

> Is there anything like an exhaustive definition that finally and once and for all sets our mind at rest? 'But are there not exact definitions at least in science?' Let's see. The notion of gold seems to be defined with absolute precision, say by the spectrum of gold with its characteristic lines. Now what would you say if a substance was discovered that looked like gold, satisfied all the chemical tests for gold, whilst it emitted a new sort of radiation? 'But such things do not happen'. Quite so; but they **might** happen, and that is enough to show that we can never exclude altogether the possibility of some unforeseen situation arising in which we shall have to modify our definition. Try as we may, no concept is limited in such a way that there is no room for any doubt. We introduce a concept and limit it in **some** directions; for instance, we define gold in contrast to some other metals such as alloys. This suffices for our present needs, and we do not probe any farther. We tend to **overlook** the fact that there are always other directions in which the concept has not been defined. And if we did, we could easily imagine conditions which would necessitate new limitations. In short, it is not possible to define a concept like gold with absolute precision, that is in such a way that every nook and cranny is blocked against entry of doubt. [35]

Vagueness

Waismann described words such as *gold* as being of an 'open texture'. He distinguishes carefully between 'open texture' and 'vagueness':

> A word which is actually used in a fluctuating way (such as *heap* or *pink*) is said to be vague; a term like *gold*, though its actual use may not be vague, is non-exhaustive or of an open texture in that we can never fill up all the possible gaps through which a doubt may seep In. [35]

In using many apparently reliable technical terms, we are troubled by their 'vagueness'. Consider such terms as *liquid*, or *at an elevated temperature*. We cannot tell to what temperature the term *elevated* applies. Presumably, minus 50°C and plus 500°C are outside the limits of this term; but just what **is** the central temperature that is *elevated*, and what ranges on either side of it represent more or less elevated temperatures? What is an *accurate* measurement, a *stable* flow-rate, or an *efficient* reaction?

Vagueness is not necessarily damaging. Many of our words are essentially generic, and usefully so. General words such as *vehicle*, *metal* or *reaction* are often properly used when a writer does not want to specify a *three-wheeled car*, *iron* or *esterification*. Our judgement of the suitability of the words must depend on whether or not the receivers are able to take from them as much information as the writer intended.

Vagueness is not only an inherent quality of many words. It is, as Willard Quine points out, also a natural consequence of the basic mechanism of word learning. We learn many of our words by observing the customary verbal response to certain stimulations. We learn by observation, imitation or instruction to use certain terms to describe certain objects or states, reactions or relationships. Our 'reward' for making the right response, or using the right word, is observably successful communication with others. Our difficulties arise when we are not entirely certain of the proper verbal signal to use in a given situation; our understanding of the term has 'fuzzy edges'.

> The penumbral objects of a vague term are the objects whose similarity to ones for which the verbal response has been rewarded is relatively slight. Or, the learning process being an implicit induction on the subject's part regarding society's usage, the penumbral cases are the cases for which that induction is most inconclusive for want of evidence. The evidence is not there to be gathered, society's members having themselves had to accept similarly fuzzy edges when they were learning. Such is the inevitability of vagueness on the part of terms learned in the primitive way; and it tends to carry over to other terms defined on the basis of these. [37]

Regrettably, though I suspect inevitably, people in science, engineering and computing, especially those interpreting the ideas and the information of other writers or teachers, often find themselves learning words in the 'primitive' way; and many technical terms are held in their understanding by means of other words which are themselves only dimly understood. There is little we can do other than try constantly to bring as many terms as possible out from the 'penumbra' into the brighter light of fuller understanding.

In other words, we have to realize that we can rarely achieve absolute precision of meaning such as would satisfy the philosophers. The best we can achieve is what (to use Waismann's expression) will suffice for our present needs, making it unnecessary to probe farther: and that must be interpreted as what will suffice in a given context to convey our information to our readers without requiring from them more than a normal commitment to disambiguation.

What will suffice

I hope this interpretation of 'what will suffice' would satisfy the text-book writers I quoted at the beginning, and Flood and Waismann. It is not question-begging or a shrugging-off of responsibility, for it does not suggest that we should go ignorant of the causes of our semantic difficulties. We must certainly be alert to the problems attached to establishing the meaning of *meaning*; but we should acknowledge that it is easier to demonstrate the incompleteness of existing theories of meaning than to devise a new theory that can be applied in all situations; and meanwhile we have to get on with communication as best we can! We can do so with reasonable success if we take care.

In my view, therefore, writers who seek advice on overcoming ambiguity must be made aware of semantic problems, but at the same time encouraged to adopt a pragmatic attitude. They must learn to identify what extent of knowledge, what range of vocabulary, what capacity to disambiguate, they can expect from various audiences. They will be in for some surprises. For example, they can expect three in every ten British first-year undergraduates in applied science departments to be unable to define *synthesis* [37]. At a more general level, they can expect four in every ten of the fifteen-year-olds leaving British secondary schools to be defeated by *circuit* [38]. But facts such as these will provide a useful balance to abstract semantic studies.

I should like to prescribe a compulsory piece of training for all embryo scientists, engineers, and computing specialists. They should be required to write a description of a start-up procedure or a set of operating instructions, or a routine to be followed in fault-finding. They should then be obliged to watch silently while an operator or maintainer tried to do the job solely in accord with the written instructions. This, I am sure, would teach as much about the right things to consider in avoiding ambiguity as several hours' reflection on abstruse aspects of semantics or philosophy.

Texts used in the survey made with the co-operation of the Institution of Chemical Engineers

Explanatory notes

Six different versions of part of an engineering report are enclosed. You are invited to show on the accompanying card which of the versions you find most comfortable to read, easiest to grasp and simplest to digest. Please add any comments you would like to make on why you prefer some versions and reject others, and return the card to The Institution of Chemical Engineers.

The versions should be judged as part of a joint-authorship paper on *Liquid Flow in Packed Columns*. The paper describes the spread of liquid through a bed of randomly dumped packings in terms of rivulets, each of which has a stable but randomly oriented path through the bed.

Version T

In all the experiments in the present work it was found that the distributions were time-independent (this has been noted by most previous workers on liquid distribution) and this tends to indicate the existence of a stable flow pattern in the packed bed. It is thought to be of interest to determine to what extent this pattern of flow is dependent on bed structure and to what extent it is dependent on movement of a purely random nature of the initial liquid particles producing wetted paths through the packing. The effect on any given arrangement of packing of complete overload of the bed by high liquid flowrate (pre-flooding) might be predicted to be the alteration of a flow pattern, dependent upon which of the paths through the packing experienced wetting, whereas the effect of re-packing or 'stirring' the bed will be changes in the structure of the bed. Liquid spread measurements utilizing salt tracer techniques were carried out both prior to and subsequent to either pre-flooding of the packed bed or effecting changes in the structure of the bed by stirring (½ in. rings were used), other conditions, packed height and liquid rate, being kept constant. It was apparent from the results that both pre-flooding and bed-stirring exerted a considerable influence on the stability of the flow pattern but that re-packing exerted a substantially higher degree of influence than pre-flooding, hence indicating that bed structure is the factor of greatest importance in the determination of flow pattern stability.

Version R

All the experiments showed the distributions to be time-independent. This has been noted by most previous workers on liquid distribution. It indicates the existence of a stable flow pattern in the packed bed. This flow pattern may be dependent on two factors. The first is bed structure, the second is purely random movement of the initial liquid particles producing wetted paths through the packing. It is of interest to determine to what extent flow pattern is dependent on these two factors. For a given arrangement of packing it could be expected that complete overloading of the bed by a high liquid flow-rate (pre-flooding) would alter a flow pattern. For the same arrangement, re-packing or 'stirring' the bed would be expected to cause changes in the structure of the bed. Liquid-spread measurements were undertaken to test this. They were carried out before and after either pre-flooding the packed bed or changing the bed structure by stirring. Half-inch rings were utilized for the latter purpose. Other conditions, namely packing height and liquid rate, were kept constant. Salt tracer experiments were used. Both pre-flooding and stirring influenced the stable flow pattern. The effect of re-packing, however, was substantially greater than that of pre-flooding. This means that stable flow pattern is determined mainly by bed structure.

Version Y

In all the experiments it was found that the distributions were time-independent (this has been noted by most previous workers on liquid distribution), and this indicates the existence of a stable flow pattern in the packed bed. It is of interest to determine to what extent this flow pattern is dependent on bed structure and to what extent it is depend-ent on a purely random movement of the initial liquid particles produc-ing wetted paths through the packing. For a given arrangement of packing, completely overloading the bed with a high liquid flow-rate (pre-flooding) might be expected to alter a flow pattern, depending on which paths through the packing are wetted, whereas re-packing, or 'stirring' the bed, will cause changes in the structure of the bed. Liquid-spread measurements were made before and also after either pre-flooding the packed bed or changing the structure of the bed by stirring (½ in. rings were used), other conditions, packed height and liquid rate, being kept constant. Salt tracer experiments were used. From the results of the experiments it can be seen that both pre-flooding and stirring the bed influence the stable flow pattern, but that the effect of repacking is much greater than that of pre-flooding. That is, the stable flow pattern is determined mainly by the structure of the bed.

Version H

All our experiments showed that the distributions did not depend on time. This has been shown by most previous work on liquid distribution, and it suggests that the flow pattern in the packed bed is stable.

We thought it would be interesting to know to what extent this flow pattern depends on bed structure and to what extent it depends on the random movement of the initial liquid particles making wetted paths through the packing. In any given packing arrangement, re-packing or 'stirring' might be expected to change the structure of the bed; complete over-loading of the bed with a high liquid flow-rate (pre-flooding) might be expected to alter a flow pattern, depending on which paths through the pattern were wetted.

So we used salt tracer methods to measure liquid spread before and after stirring (using ½ in. rings) and before and after pre-flooding. Other conditions – packed height and liquid rate – were kept constant. Both stirring and pre-flooding affected the stable flow pattern, but stirring had much greater effect than pre-flooding: obviously, the structure of the bed is mainly responsible for the stable flow pattern.

Version F

Time-independence of distributions, indicative of stable flow pattern in the packed bed, occurred in all experiments, confirming most previous liquid distribution work. The degree of dependence of this pattern on bed structure and/or on production of wetted paths through the packing by randomly moving particles of the initial liquid is of interest, as in any specified packing arrangement, complete bed overload by high liquid flow-rate (pre-flooding) might be expected to result in alteration of a flow pattern, dependent on which paths through the packing were wetted, while changes in bed structure might result from re-packing or 'stirring' the bed. Liquid-spread measurements by salt tracer techniques were made before and after either packed-bed pre-flooding or structure-changing by stirring (utilizing ½ in. rings); other conditions, packed height and liquid rate, being maintained constant. The effect of re-packing on flow pattern stability was far in excess of that of pre-flooding. Hence flow pattern stability is determined mainly by bed structure.

Version B

In all our experiments we found that our distributions were not dependent on time. This has been noted by the majority of our predecessors in work in the liquid distribution field, and we believe that this indicates the existence of a stable flow pattern in the packed bed.

We felt it would be of considerable interest if we were able to demonstrate in our experiments the extent to which this pattern depends on bed structure and to what extent it depends on a purely random movement of the initial liquid particles producing wetted paths through the packing. We might reasonably hypothesize that for any given packing arrangement, complete overloading of the bed by a high liquid flow-rate (known as pre-flooding) would alter our flow pattern, depending on which paths through the packing were wetted. We would expect, on the other hand, that re-packing or 'stirring' the bed would bring about changes in the fundamental structure of the bed.

We undertook measurements of liquid spread before and following either pre-flooding the packed bed or changing the structure of the bed by stirring (which we effected by means of ½ in. rings). Other conditions, viz. packed height and liquid rate, we maintained at a constant level. Salt tracer techniques were our choice for the experiments.

Our results reveal that both pre-flooding and stirring the bed exert a marked influence on the stable flow pattern; nonetheless, the effect of re-packing is of far greater magnitude. We must conclude, therefore, that the stable flow pattern is determined mainly by the structure of the bed.

Text used in the survey made with the co-operation of the British Ecological Society

Explanatory notes

Members of the British Ecological Society are invited to co-operate in an experiment aimed at establishing what style is most readable and acceptable for scientific reports and papers.

Six different versions of part of a scientific paper are enclosed. You are invited to show on the accompanying card which of the versions you find most comfortable to read, easiest to grasp and simplest to digest. Please add any comments you would like to make on why you prefer some versions and reject others, and return the card.

The versions should be judged as part of a joint-authorship paper for a professional journal such as *The Journal of Applied Ecology*. The paper describes work aimed at discovering which types of pasture are preferred for grazing by wigeon. Full nomenclature for the wigeon (*Anas penelope* L.) and for the species of grasses (*Festuca rubra* L., *Puccinellia maritima* (Huds) Parl, and *Agrostis stolonifera* L.) is given in the title of the paper.

There is no set order in which the versions should be read. Indeed, it is desirable that there should be as much variety in order of reading as possible. To achieve this, I suggest that each readers starts with the letter nearest to the initial letter of his or her surname.

For the sake of realism, the versions are based on genuine subject-matter reported in a published paper. I would stress, however, that none of these versions is the responsibility of the authors of the original paper!

Version S

Bird Island saltings comprise three abundant species of grass. The species are *Festuca rubra* (red fescue), *Puccinellia maritima* (salt-marsh grass) and *Agrostis stolonifera* (creeping bent grass). For the purpose of this study, two major types of sward were distinguished subjectively. These were areas dominated by *Festuca rubra* and those by *Puccinellia maritima*. Of the area studied, approximately 59% was swards dominated by *Festuca rubra*. These swards also contained varying proportions of *Agrostis stolonifera* and a little *Puccinellia maritima*. Swards in the other 41% of the area were dominated by *Puccinellia maritima*. These swards also contained varying proportions of *Agrostis stolonifera* and a little *Festuca rubra*.

Observations were carried out from an observation tower overlooking the Island in the winter of 1970–71. It was observed that the degree of disturbance due to visitors to the area was of a sufficient order to inhibit wigeon from frequenting the Island during daylight hours. However, droppings were observed on the Island, and this together with limited observations carried out at sunrise, sunset and on moonlit nights suggested the return of wigeon to graze at night.

It was decided that a count should be made of the numbers of wigeon feeding on the two main types of sward represented on the Island. This would enable a comparative assessment of the feeding preferences of the wigeon to be made. Owen (1971) suggested that in view of the fact that European white-fronted geese (*Anser anser albifrons* (Scopoli)) defecate every 3½ minutes, a droppings count is a sensitive method of estimating grazing usage. The same basic assumption was used in the determination of feeding preferences in this work. It was assumed that the wigeon's quantitative defecation would be in proportion to the length of time spent grazing.

Counts were made in six plots in each of the two swards. Inter-count intervals were dictated by times when submergence of the Island by high spring tides made counting impossible. To avoid removal of the droppings by the tide, the first count was undertaken just prior to flooding. After each count, droppings were removed or trodden into the ground. Hence the possibility of duplication during ensuing counts was avoided. Each plot size was 10 × 3 m. Posts used as markers were small to minimize disturbance to the wigeon by alien objects on the feeding ground.

Selection of the six plots in each sward was based on an initial visual differentiation between the two swards. Due to morphological differences, the naked eye could readily detect differences between the swards. The swards in which the plots were placed were later compared by a $1/20\,m^2$ ($25 \times 20\,cm$) quadrat sampling technique. Percentage cover estimates were made for the three grass species in twenty randomly selected quadrats in each plot. To test for changes to the sward consequent upon wigeon grazing, the percentage frequency occurrence of various grass blade lengths of the species in a *Puccinellia/ Agrostis* plot was determined. Two determinations were carried out at a one month interval, at times when maximum wigeon numbers were present. A full description of the point intercept method utilized in the determinations is given in Appendix 1.

Version B

The saltings on Bird Island have three abundant species of grass, namely *Festuca rubra* (red fescue), *Puccinellia maritima* (salt-marsh grass) and *Agrostis stolonifera* (creeping bent grass). Approximately 59% of the study area was covered by swards in which *Festuca rubra* predominated with varying proportions of *Agrostis stolonifera* and a little *Puccinellia maritima*. The remaining area consisted mainly of *Puccinellia maritima* but with a little *Festuca rubra,* and again with varying proportions of *Agrostis stolonifera.* For the purpose of this study two major sward types were distinguished subjectively, namely areas dominated by *Festuca rubra* and those by *Puccinellia maritima.*

In order to carry out a comparative assessment of the feeding preferences of the wigeon, it was decided that a count should be made of the numbers feeding on the two main types of sward represented on the Island. Owen (1971) suggested that as the European white-fronted geese (*Anser anser albifrons* (Scopoli)) defecate every 3½ minutes, counting the droppings is a sensitive method for estimating grazing usage. The same basic presumption, namely that the wigeon would defecate quantitatively in proportion to the length of time spent grazing, was used in the determination of feeding preferences in this work. Equal numbers of plots were marked out in both swards. Observations carried out from an observation tower overlooking the Island in the winter of 1970–71, however, showed that the degree of disturbance caused by visitors to the area was sufficient to keep the wigeon off the Island during the day. The presence of droppings on the Island and limited observations carried out at sunrise, sunset and moonlit nights indicated that the wigeon were returning to graze at night.

As this study was intended to compare the preference of wigeon for the two types of sward on Bird Island, the number of droppings was counted on three and four occasions during the period November–March 1971/72 and 1972/73 respectively in six plots in each of the two swards. The interval between counting was governed by the times when the Island was covered at high spring tides. The first count was just before flooding, as otherwise the droppings would have been washed away. The droppings were removed or trodden into the ground following each count in order to prevent duplication during ensuing counts. Each plot was 10 × 3 m and was marked with small posts in order to reduce disturbance of the wigeon by the presence of alien objects on their feeding ground.

The selection of the sites of the six plots in each sward was made on an initial visual differentiation between the two swards which owing to their different morphology were quite distinct to the naked eye. The swards in which the plots were placed were later compared by a $1/20\,m^2$ (25 × 20 cm) quadrat sampling technique, in which percentage cover estimates were made for the three grass species in twenty random quadrats in each plot. In order to test for changes to the sward as a result of wigeon grazing, the percentage frequency occurrence of the various blade lengths of the grass species in a *Pucccinellia/Agrostis* plot was determined by a point intercept method on two occasions at a month's interval, when maximum numbers of wigeon were present. This intercept method is described in full in Appendix 1.

Version M

Three sorts of grass grow abundantly on Bird Island. They are red fescue (*Festuca rubra*), salt-marsh grass (*Puccinellia maritima*) and creeping bent grass (*Agrostis stolonifera*). About 59% of our study area was covered by swards which were mainly red fescue, with varying amounts of creeping bent grass and a little salt-marsh grass. The rest was mainly salt-marsh grass, with varying amounts of creeping bent grass and a little red fescue. For the purpose of our study, we subjectively distinguished between two major sward types: one dominated by the fescue and one dominated by the marsh grass.

As it was our intention to assess which types of grass the wigeon preferred for feeding, we decided to count the numbers grazing on our two main types of sward. We knew that Owen, in 1971, had suggested that as European white-fronted geese (*Anser anser albifrons* (Scopoli)) defecate every 3½ minutes, a count of the droppings gives you a good idea of the extent to which the birds are using the grazing. So we worked on the same assumption: that is, we took it that we could base our assessment of feeding preferences on the idea that the number of wigeon droppings would be proportional to the time they spent grazing.

We marked out the same number of plots in each sward. We knew from observations we had made in the winter of 1970–71 that visitors to the area scared the birds away from the Island during the day. But we also knew that the birds were coming back to graze at night, because we found droppings on the Island, and we kept watch a few times at sunrise, sunset, and on moonlit nights.

Our idea being to see which types of sward the wigeon preferred, we counted the number of droppings in six plots in each of the two swards on seven occasions in all – three in November–March 1971–72 and four in November–March 1972–73. Of course, we could not count when the Island was under water during the high spring tides. The first count was just before the island was flooded, when the droppings would have been washed away. After each count, we always took the droppings right away or else trampled them well into the ground, so that we would not count the same droppings again during the next count. The Plots were 10 × 3 m in size, and were marked with posts – **small** posts, so that the wigeon would be disturbed as little as possible by strange things in their feeding grounds.

At first, we chose the sites for the six plots in each of the two swards simply by looking at them: they were plainly different in form. Later, we compared them by a 25 × 20 cm quadrat sampling technique, to get an estimate of the percentage cover for our three grass species. This was done in twenty random quadrats in each plot. To test for changes in the sward after the wigeon had grazed, we used a point intercept method to check the percentage frequency of the lengths of blades of grass in a salt-marsh-dominated plot. We did this twice, with a month in between, at times when the largest number of wigeon were present. Full details of our intercept method are in Appendix 1.

Version Y

Three species of grasses grow abundantly on Bird Island saltings: red fescue, salt-marsh grass and creeping bent grass (*Festuca rubra, Puccinellia maritima* and *Agrostis stolonifera*). For our study, we distinguished subjectively between two main types of sward: swards dominated by red fescue and swards dominated by salt-marsh grass. Approximately 59% of the study area was dominated by red fescue, with a little salt-marsh grass and varying proportions of bent grass. The remaining 41% was dominated by salt-marsh grass, with a little red fescue and varying proportions of bent grass.

In the winter of 1970–71, observations from a tower overlooking the island showed that visitors to the area caused enough disturbance to keep the wigeon off the island during the day. However, the presence of droppings on the island, and a few observations at sunrise, sunset and on moonlit nights indicated that the wigeon were returning to graze at night.

Our aim was to assess feeding preferences by counting the numbers of wigeon feeding on the two types of sward. In 1971, Owen suggested[1] that as European white-fronted geese (*Anser anser albifrons* (Scopoli)) defecate every 3½ minutes, a count of their droppings gives a sensitive estimate of the extent of their grazing. We decided to make the same assumption: that the number of droppings would show how long the wigeon spent grazing, and thus which grass they preferred. Accordingly, we marked out equal numbers of plots in both types of swards. Each plot was 10 × 3 m. The posts marking the plots were small, to minimize disturbance of the birds by strange objects on their feeding grounds.

The droppings in six plots in each sward were counted on three occasions in November–March 1971–72 and on four occasions in November–March 1972–73. The intervals between counts were governed by the high spring tides, which covered the island. The first count was just before flooding, which would have washed away the droppings. After each count, the droppings were removed or trodden into the ground, to prevent duplication in subsequent counts.

As the swards were visibly different in form, the sites of the six plots in each sward were selected subjectively at first. Later, the swards containing the plots were compared by a 25 × 20 cm quadrat sampling technique, in which percentage cover was estimated for the three species of grasses. Twenty quadrats were sampled at random in each plot.

To test whether grazing by wigeon changed the sward, we calculated the percentage frequency of blade lengths of the grasses in a plot dominated by salt-marsh grass. The calculation was made on two occasions, one month apart, when the maximum number of wigeon were present. The point intercept method used is described fully in Appendix 1.

Version F

Three grass species, *Festuca rubra* (red fescue), *Puccinellia maritima* (salt-marsh grass), and *Agrostis stolonifera* (creeping bent grass) occur abundantly on Bird Island saltings. For this study, two major sward types (*Festuca*-dominated, containing varying amounts of *Agrostis stolonifera* and a small quantity of *Puccinellia maritima*, and *Puccinellia*-dominated, containing varying amounts of *Agrostis stolonifera* and a small quantity of *Festuca rubra*) were distinguished subjectively, comprising 59% and 41% of the study area respectively.

Observations from an observation tower overlooking the Island in winter 1970–71 indicated an incidence of disturbance due to visitors sufficient to cause diurnal absence of wigeon from the Island. However, limited observations at sunrise, sunset, and on moonlit nights, and presence of droppings on the island suggested nocturnal return and resumption of grazing.

Comparative assessment of wigeon feeding preferences by determination of numbers feeding on the Island's main sward types was carried out by counting of droppings. Following Owen's (1971) theory that since European white-fronted geese (*Anser anser albifrons* (Scopoli)) defecate at 3½ minute intervals, a sensitive measure of grazing usage may be obtained from a droppings count, it was decided to accept number of wigeon droppings as a function of time spent grazing. In both swards, demarcation of equal numbers of 10 × 3 m plots was accomplished by utilization of small posts to reduce wigeon disturbance effects due to presence of alien objects on the feeding ground.

To compare feeding preferences, a count was made of numbers of droppings in six plots in each of the two sward types on three and four occasions during November–March 1971–72 and 1972–73 respectively, the interval between counts being governed by immersion of the Island at the time of high spring tides. Thus the initial count took place immediately prior to flooding to avoid removal of droppings by water.

Duplication during ensuing counts was prevented by removal of droppings or treading into the ground.

Site selection of the six plots in each sward was based on an initial
visual differentiation between the sward types, whose heterogeneity in
respect of morphology was macroscopically apparent. A comparison of
swards in which plots were located was however later performed
utilizing a 25 × 20 cm quadrat sampling technique, enabling percentage
cover estimation for the three grass species in twenty randomly
selected quadrats in each plot. As a test for sward modifications consequent upon wigeon grazing, a point intercept method was employed
(for full details see Appendix 1) on two occasions at an interval of one
month, at peak wigeon population levels, to determine percentage
frequency occurrence of various blade lengths of grass species in a
Puccinellia/Agrostis plot.

Version R

In the saltings in question three abundant grass species occur, namely
Festuca rubra (red fescue), *Puccinellia maritima* (salt-marsh grass), and
Agrostis stolonifera (creeping bent grass). Of the area concerned in the
study, of the order of 59% was covered by swards in which grass of
the *Festuca rubra* type was predominant, with proportions of *Agrostis
stolonifera* at varying levels, and also a small amount of *Puccinellia
maritima*, whilst in the remaining 41% (approximately) *Puccinellia
maritima* was predominant, with a small amount of *Festuca rubra*, and
again proportions of *Agrostis stolonifera* at varying levels. In the current
study, two principal types of sward were distinguished initially for
experimental purposes on a subjective basis, namely areas dominated
by *Festuca rubra* and those dominated by *Puccinellia maritima*.

It was thought that a comparative assessment of the feeding preferences of the wigeon could be achieved by observation of the numbers
engaged on feeding on the two main types of sward. A previous
worker (Owen, 1971) had suggested that as the European white-
fronted geese (*Anser anser albifrons* (Scopoli)) defecate at 3½ minute
intervals, grazing usage is capable of accurate measurement by the
method of counting droppings. A similar premiss was postulated in
this work: that a quantitative measure of defecation would be directly
proportional to the total length of time spent grazing by the wigeon,
and from this feeding preferences could be determined. Within each of
the principal types of sward distinguished on the Island, equal numbers of plots were identified with markers. From observations carried
out from an observation tower located in a position overlooking the
Island in the winter of 1970–71, it was apparent that the degree of

disturbance caused by the presence of visitors to the area was sufficient to cause the wigeon to refrain from frequenting the Island in the course of the day. On the other hand, from the presence of droppings on the Island and from a limited number of observations undertaken at sunrise, sunset and moonlit nights it was apparent that the birds were returning at night to resume their grazing activity.

As it was the intention of the current study to perform a comparison of the preferences demonstrated by the wigeon between the two principal types of sward on Bird Island, a count was made of the number of droppings on three and four occasions during the period between November–March 1971–72 and November–March 1972–73 respectively in each of six plots in each of the two sward types. The factor governing the interval between counting was the times during which immersion of the Island by high level spring tides was complete. The initial count was carried out immediately prior to flooding as otherwise removal of the droppings by washing action of the water would have taken place. Following each count, droppings were removed or trodden firmly into the ground, thus preventing duplication during counts on subsequent occasions. Each plot was $10 \times 3\,\mathrm{m}$ in area and was marked with small posts in order to diminish the disturbance to the wigeon occasioned by the presence of alien objects in their customary feeding ground.

Site selection for the six plots in each sward types was based on an initial differentiation between the two swards made visually, the swards being quite distinct to the unaided eye due to their morphological distinctiveness. A comparison of the swards in which the plots were placed was subsequently performed by means of a $1/20\ \mathrm{m}^2$ ($25 \times 20\,\mathrm{cm}$) quadrat sampling technique, in which percentage cover estimates were made in twenty quadrats chosen at random in each plot for each of the three grass species under review. To identify changes in the sward brought about by the grazing of the wigeon a test was carried out to determine by a point intercept method the percentage frequency occurrence of blades of varying lengths of the grass species occurring in a *Puccinellia/Agrostis* plot on two occasions separated by a one month interval, when maximum numbers of wigeon were present. Full details of the intercept method concerned are shown in Appendix 1.

Texts used in the survey made with the co-operation of the Biochemical Society

What is good style for scientific writing?

Members of the Biochemical Society are invited to co-operate in an experiment aimed at establishing what style is most readable and acceptable for scientific reports and papers.

Six different versions of part of a scientific paper follow. You are invited to show on the attached slip which of the versions you find most comfortable to read, easiest to grasp and simplest to digest. Please add any comments you would like to make on why you prefer some versions and reject others, and return the slip.

The versions should be judged as part of a joint-authorship paper for a professional journal such as the *Biochemical Journal*. The paper describes an investigation of how the secretion of growth hormone from anterior pituitary of rats is affected by cyclic AMP. The extract is taken from the introductory section of the paper.

There is no set order in which the versions should be read. Indeed, it is desirable that there should be as much variety in order of reading as possible. To achieve this, each reader is asked to start with the letter nearest to the initial letter of his or her surname.

For the sake of realism, the versions are based on genuine subject-matter reported in a published paper. However, none of these versions is the responsibility of the authors of the original paper!

Version A

This study was to devise a simple system for further investigation *in vitro* of control of growth hormone secretion. It was to be based on measurement of growth hormone release rate from isolated fragments of rat anterior pituitary. Rat pituitary was chosen because of its ready availability in the fresh state. It is also relatively small. This allows fragments to be obtained with minimal tissue damage. The investigation used the double antibody radioimmunoassay technique recently developed in the Biochemistry Department of the University of Wessex (Black & White, 1970). This procedure is rapid, sensitive and specific for rat growth hormone assay. It has been shown to be valid when applied to the determination of the hormone released into media in which rat anterior pituitary fragments have previously been incubated (Lavender, Greenhedge, Hawthorne & Berry, 1972).

A preliminary account of some of the findings reported here has previously been published (Brown & Green 1973). That account included a demonstration of a direct effect of dibutyryl cyclic AMP in stimulating growth hormone secretion *in vitro*. The present paper shows that rat pituitary fragment preparation is a reproducible and sensitive system for *in vitro* investigation of regulation of growth hormone release. Also, it is demonstrated that it is possible to distinguish between true secretion and hormone leakage. Hormone leakage, it is presumed, occurs due to cell damage during isolation of gland fragments. The system has been employed to study growth hormone secretion. It has also been used in the investigation of the metabolic requirements of the gland when active secretion is occurring in response to stimulation by cyclic AMP.

Version G

We set out to produce in this work a simple *in vitro* system which we could use to look further into the ways in which secretion of growth hormone can be controlled. We based our method on measurement of the rate at which growth hormone is released from isolated fragments of anterior pituitary from rats. We chose to use rat pituitary because it is easy to get fresh supplies and it is relatively small. This meant we could obtain fragments with only a little tissue damage.

In our work, we used the rapid, sensitive and specific double-antibody technique for radioimmunoassay of growth hormone in rats, developed recently in the University of Wessex [1]. We knew it had been shown to be valid when used to check on the amounts of hormone release into media in which fragments of rat anterior pituitary had been incubated previously [2].

We have published some of our findings previously [3], including our demonstration of the direct effect dibutyryl cyclic AMP has on the stimulation of growth hormone secretion *in vitro*. In this paper, we present our evidence to support the concept that rat pituitary fragment preparation gives a reproducible and sensitive system for *in vitro* investigation of the regulation of growth hormone release. Also, we describe how, by using it, we have been able to distinguish between hormone leakage, which we presume occurs because of damage to cells during isolation of gland fragments, and true secretion. We have used the system to study growth hormone secretion, and we have also used it to investigate what a pituitary gland's metabolic requirements are when it is in an actively secreting state, responding to stimuli from cyclic AMP.

Version O

This study aimed at devising a simple *in vitro* system for further growth hormone secretion control mechanism investigations, based on growth hormone release rate measurements from isolated rate anterior pituitary fragments. Choice of rat pituitary was due to its relatively small size, thus permitting fragment section with minimal tissue damage, and ready availability in the fresh state. The University of Wessex's recently developed rapid, sensitive, specific, double-antibody radioimmunoassay technique (Black & White, 1970), whose validity has been demonstrated in application to determination of hormone release into media previously utilized for incubation of rat anterior pituitary fragments (Lavender, Greenhedge, Hawthorne & Berry, 1972) was used in these investigations. Preliminary findings, including demonstration of a direct effect of dibutyryl cyclic AMP on growth hormone secretion stimulation *in vitro*, have been published (Brown & Green, 1973). This paper shows rat pituitary fragment preparation to provide a reproducible and sensitive system for *in vitro* investigation of growth hormone release regulation, usable to distinguish between hormone leakage (occurring presumably from cell damage during gland fragment isolation) and true secretion. Study of growth hormone secretion and investigation of the metabolic conditions necessary for cyclic AMP-stimulated secretory processes to occur actively in the gland have been performed with this system.

Version W

It was the purpose of the present study to devise a simple system *in vitro*, based on measurement of the rate of growth hormone release from isolated fragments of rat anterior pituitary, which would be suitable for the further investigation of the mechanisms by which growth hormone secretion may be controlled. The choice of rat pituitary was made for these studies on grounds of its ready availability in the fresh state and its relatively small size which allows fragments to be obtained with minimal tissue damage. The investigations described below were carried out using the rapid, sensitive and specific double-antibody radioimmunoassay technique for rat growth hormone measurement recently developed in the Biochemistry Department of the University of Wessex (Black & White, 1970) which has been shown to be valid when applied to the determination of the hormone release into media in which rat anterior pituitary fragments have previously been incubated (Lavender, Greenhedge, Hawthorne & Berry, 1972).

A preliminary account of some of the findings reported here, including the demonstration of a direct effect of dibutyryl cyclic AMP in stimulating growth hormone secretion *in vitro*, has previously been published (Brown & Green, 1973). In the present paper, evidence is presented to show that the rat pituitary fragment preparation provides a reproducible and sensitive system for the investigation *in vitro* of the regulation of growth hormone release and that, by its use, it is possible to distinguish between hormone leakage, occurring presumably from cells damaged during isolation of the gland fragments, and true secretion. The system has been employed to study growth hormone secretion and to investigate the metabolic requirements exhibited by the gland when it is stimulated into an actively secreting state by cyclic AMP.

Version P

The purpose of the study reported here was the establishment of an unsophisticated *in vitro* system, based on measurement of the rate of release of growth hormone from isolated fragments of anterior pituitary from rats, which would provide a suitable technique for further elucidation of the mechanisms by which regulation of the growth hormone secretory process in the cells of the anterior pituitary may be achieved. The choice of rat anterior pituitary for these studies was made by virtue of the ready availability of this gland in the fresh state and its relatively small size which permits fragments being obtained with minimal damage to the tissues involved.

The investigations described in the following pages were carried out utilizing the rapid, sensitive, specific, double-antibody radioimmunoassay technique for rat growth hormone measurement recently developed in the Biochemistry Department of the University of Wessex (Black & White, 1970). That such a process has validity has been demonstrated by its application to the determination of hormone release into media in which incubation of rat anterior pituitary fragments has previously been carried out (Lavender, Greenhedge, Hawthorne & Berry, 1972).

A preliminary publication (Brown & Green, 1973) presented an account of some of the experimental findings reported below in this paper, including evidence demonstrating the direct role of dibutyryl cyclic AMP in the stimulation of secretion of growth hormone *in vitro*.

The present paper presents evidence to demonstrate that a sensitive and reproducible system for the *in vitro* investigation of the growth hormone release regulatory mechanisms is provided by the rat pituitary fragment preparation. Utilizing this system, distinction is possible between hormone leakage – occurring, it is presumed, due to cell damage during isolation of gland fragments – and true secretion. Studies of growth hormone secretion have been carried out employing this system, and also investigations of the metabolic requirements exhibited by the gland when stimulation to an actively secreting state by cyclic AMP is occurring.

Version J

Our aim was to devise a simple system for further *in vitro* investigation of ways of controlling secretion of growth hormone. The investigation was to be based on measuring the rate at which growth hormone is released from isolated fragments of anterior pituitary from rats. We chose rat pituitary because it is relatively small, so fragments can be obtained with only slight tissue damage. Also, fresh pituitary is easy to get. In the investigation, we used the rapid, sensitive and specific technique developed at the University of Wessex[1] for double-antibody radioimmunoassay of growth hormone from rats. This assay has been shown[2] to be valid when used to calculate how much hormone is released into media in which fragments of anterior pituitary from rats have been incubated.

Some of our findings, including the direct stimulation of growth hormone secretion by dibutyryl cyclic AMP *in vitro*, have been published previously[3]. In this paper, we show that a preparation of fragments from rat pituitaries is a reproducible and sensitive system for investigating *in vitro* how the release of growth hormone can be regulated. Using this preparation, it is possible to distinguish between true secretion and hormone leakage (which presumably occurs because cells are damaged when fragments of gland are isolated). We have used the preparation to study the secretion of growth hormone, and to investigate the metabolic conditions required if the pituitary is to secrete actively in response to cyclic AMP.

Results and comments from the surveys of style for reports and papers

Which style do readers prefer for scientific and technical reports and papers? In all three surveys, the respondents voted clearly in favour of direct, active writing, with a minimum of specialist vocabulary, a judicious mixture of personal and impersonal constructions, short and uncomplicated sentences, and liberal paragraphing. In all three surveys, a reasonably well written version in 'traditional' passive, impersonal style came a poor second. The majorities were 28%, 41% and 18.5%. The full results were as shown in Tables 1–3 pages 191–3.

I readily acknowledge a point made by many of the readers who responded to my surveys. They pointed out that my samples of readers' views would be biased – that I should get replies from people who care about skilful writing but not from those who do not. I acknowledge that weakness; it is inherent in any surveys that rely on voluntary responses, especially voluntary responses by post. But I believe it is better to have these samples than to have no real evidence about scientists' opinions on writing. And I would suggest that the opinions of those who care about skilful writing are the ones we should value.

It is important to acknowledge, too, that we are unlikely ever to get unanimity about the readability of writing, because judgement is entirely subjective. I must repeat that the readability of a text is affected by many factors besides the choice of language. Obviously, readers' background knowledge of the subject affects their capacity to comprehend what is being said. The intrinsic difficulty of the concepts being expressed is important. Readers' motivation is also significant: do they **want** to read the text, or are they being forced to? Do they want to scrutinize the text closely, or is their aim merely to skim it superficially? To make a full analysis of readers' responses to a text, we should take into account a wide range of psychological and even physical elements in the context in which the reading is taking place.

The influence of linguistic features

This survey aimed only to show the influence of a number of linguistic features on the readability of a text. I wanted to stress the difference that can be made by choosing language that is active rather than passive, concrete rather than abstract, and 'everyday' rather than 'special'; to show that the digestibility of statements is affected by both the length and the complexity of the sentences used to express them; to show the importance of using a variety of sentence types, some personal and some impersonal, instead of artificially restricting the choice to one or the other; and to stress the importance of breaking the whole statement (that is, the whole text) into units that give readers the information they need in convenient order and in digestible 'bites'.

Table 1 Results of the survey made with the co-operation of the Institution of Chemical Engineers

Version	Description	No. of times chosen as 'best'	
H	Direct, verbs mainly active, minimum of special vocabulary, judicious use of personal and impersonal constructions, sentences of varied length but mainly short and not complex, 3 paragraphs.	706	(46%)
Y	Reasonably direct, verbs mainly passive, some roundabout phrasing to avoid personal constructions, reasonable amount of special vocabulary, sentences of varied length but some long, involved constructions, 2 paragraphs.	280	(18%)
R	Reasonably direct, verbs mainly passive, some roundabout phrasing to avoid personal constructions, moderate use of special vocabulary, sentences almost all short, consequent bitty effect, 1 paragraph.	245	(16%)
F	Reasonably direct, verbs mainly passive, much roundabout phrasing to avoid personal constructions, consequent extensive use of abstractions, much use of nouns as adjectives, sentences often long and too tightly packed with information, 1 paragraph.	133	(9%)
B	Chatty but clear, verbs mainly active, little special vocabulary, some colloquial expressions, excessive use of personal constructions, sentences of varied length but mainly short and of simple structure, 4 paragraphs.	121	(8%)
T	Roundabout and woolly, verbs mainly passive, excessive use of special vocabulary, much roundabout phrasing, sentences often long and involved, 1 paragraph.	50	(3%)
		1535	(100%)

Table 2 Results of the survey made with the co-operation of the British Ecological Society

Version	Description	No. of times chosen as 'best'	
Y	Direct, verbs mainly active, minimum of special vocabulary, judicious use of personal and impersonal constructions, sentences of varied length but mainly short and not complex, 6 paragraphs.	302	(57.4%)
B	Reasonably direct, verbs mainly passive, some roundabout phrasing to avoid personal constructions, reasonable amount of special vocabulary, sentences of varied length but mainly short and not complex, 4 paragraphs.	86	(16.4%)
F	Reasonably direct, verbs mainly passive, much roundabout phrasing to avoid personal constructions, consequent extensive use of abstractions, excessive use of special vocabulary, sentences often long and too tightly packed with information, 5 paragraphs.	56	(10.7%)
S	Reasonably direct, verbs mainly passive, some roundabout phrasing to avoid personal constructions, too much special vocabulary, much use of nouns as adjectives, sentences almost all short, consequent bitty effect, 5 paragraphs.	46	(8.7%)
M	Conversational but clear, verbs mainly active, little special vocabulary, some colloquial expressions, excessive use of personal constructions, sentences of varied length but mainly short and of simple structure, 5 paragraphs.	30	(5.7%)
R	Roundabout and woolly, verbs mainly passive, excessive use of special vocabulary, much roundabout phrasing, sentences often long and involved, 4 long paragraphs.	6	(1.1%)
		526	(100%)

Table 3 Results of the survey made with the co-operation of the Biochemical Society

Version	Description	No. of times chosen as 'best'	
J	Direct, verbs mainly active, minimum of special vocabulary, judicious use of personal and impersonal constructions, sentences of varied length but mainly short and not complex, 2 paragraphs.	283	(38.2%)
W	Reasonably direct, verbs mainly passive, some roundabout phrasing to avoid personal constructions, reasonable amount of special vocabulary, sentences of varied length but some long, involved constructions, 2 paragraphs.	146	(19.7%)
G	Conversational but clear, verbs mainly active, little special vocabulary, some colloquial expressions, excessive use of personal constructions, sentences of varied length but mainly short and of simple structure, 3 paragraphs.	137	(18.5%)
A	Reasonably direct, verbs mainly passive, some roundabout phrasing to avoid personal constructions, moderate use of special vocabulary, sentences almost all short, consequent bitty effect, 2 paragraphs.	104	(14.0%)
P	Roundabout and woolly, verbs mainly passive, moderate use of specialized vocabulary, much verbose phrasing, sentences often long and involved, 4 paragraphs.	37	(5.0%)
O	Reasonably direct, verbs mainly passive, much roundabout phrasing to avoid personal constructions, consequent extensive use of abstractions, much use of nouns as adjectives, sentences often long and too tightly packed with information, 1 paragraph.	34	(4.6%)
		741	(100%)

Clear majorities of readers agreed which tactical choices produced the best effect. Indeed, many readers expressed views about the Versions that were less polite than my thumbnail sketches in the tables of results. For example, comments on Version F of the 'wigeon' texts included:

'heavy, laborious and ponderous'
'complex sentence structure'
'sense obscured by the complicated sentence construction'
'sentences too long and complicated'
'short, but style is jargon, not very explicit and difficult to read'
'style very bad, meaning unclear'
'contains an excessive proportion of Latin derived polysyllables and apposition nouns in substitution for adjectives for either ready comprehensibility or acceptability'
'dreadful jargon'
'the deliberate use of jargon and nouns in apposition is excruciating English at its Germanic worst'
'too much use of the passive'
'clarity sacrificed for brevity'
'almost unreadable'
'only after reference to the others was comprehension complete'
'practically incomprehensible'
'a caricature'
'technical narcissism, i.e. writing for the writer's sake, not the reader's'
'the commonest style that scientists use'
'would probably be accepted for a learned journal, I am afraid to say'
'unfortunately this kind of writing gets into print on occasion'
'I've chosen F as worst mainly because the style is so common in journals'
'the booby prize...(but)...even F is reasonably lucid compared to some recent articles in the *Journal of Ecology*'

In contrast, 10.7% of the respondents in the same survey felt that Version F was the **best**:

'concise, precise and to the point'
'best: concise, logical progression of ideas, good organization makes it digestible'
'concise, fully informative, least ambiguous or vague statement'
'no surplus words'
'preferred because it is the most succinct'
'crisp, clear, easily understood'
'stands well above the others'

Plainly, judgements are made on more than linguistic features alone. The influence of custom and expectation is strong. This was demonstrated in respondents' reactions to Version M of the 'wigeon' texts. Most readers (including me) found Version M irritating and distastefully self-congratulatory. However, it is not the use but the **abuse** of personal constructions that is to blame. Too many 'we's', combined with some colloquial expressions and a general looseness of construction, led to the view that Version M was too 'chatty' and lacking the formality due in a scientific paper.

The importance of linguistic propriety

Many respondents were surprised and intrigued by their own reactions to Version M. Their views are summed up in these quotations:

'I am not sure why I object to the very personal and conversational approach in M. It is otherwise good'

'clear, but I have inbuilt bias against 1st person writing in journals'

'while version M is attractive, easily read, would it ever pass an editor?'

'this version went in best of the lot, but was far too matey to avoid irritating me'

'good television script'

'the rather chatty style of M seems to get the idea across much quicker'

'too self-consciously simple for the type of journal; fine for the layman although the meaning is clear'

'childish though easily read'

'clear but a trifle juvenile'

'is quite childish but is very easily understood'

'God preserve us from M!'

'although readable, is irritatingly personal and chatty'

'my objections to M are probably the result of a conservative reaction to the journalistic tone. The style seems more appropriate for spoken presentation but the information is presented clearly and directly'

'I suspect I did not rate (M) highly because it is an unusual way of writing scientific results'

'a bit folksy and naive, but clear and readable'

'is really unacceptable in a scientific journal but does communicate fairly economically'

'annoying for a **scientific** publication, although highly suitable for a talk or a semi-popular article'

'comes over more as an **oral** presentation'

'would be first or second preference if I were **listening** to the paper'

'this version says it all very clearly but the lapses into journalese/ American come as a shock and distract from the feeling of scientific objectivity'

'very clear, but chatty style feels wrong; excellent for verbal delivery'

'possibly alright for a talk to a natural history soc. but not for a published paper'

These repeated assessments of M as very clear but not acceptable in a formal scientific paper accentuate the weight that must be given to linguistic propriety. They express objections that have nothing to do with accuracy or clarity. They carry at least a tinge of scientific snootiness. The facts in Version M are the same as the facts in all the other versions, and there is no suggestion that those facts in themselves are unworthy of scientific attention. So it must have been the language chosen in Version M that made many readers reject it as lacking due scientific dignity.

To identify this phenomenon is not to belittle it. It is a phenomenon to be found in most communication contexts, not just in science. I have drawn attention to it here because it seems to me important that we should understand the bases of our judgements about the acceptability of attempts at communication. In discussion of Version M, we are brought face-to-face with the differences between what is acceptable expression in speech and in what is acceptable expression in writing. If we are to become skilled presenters of scientific information, we must become sensitive to such differences.

Are scientists and engineers, on average, skilled writers of reports and papers? Is there need for advice on style such as is offered in this book? Certainly, many respondents to my surveys thought there **is** a need. Let me answer with one quotation from each of these surveys of style for reports and papers:

> ...it is certainly time that biologists gave more attention (and had their attention drawn) to the material they publish. **Most** of it is so tedious, impenetrable or boring that I am sure one of the main reasons why fewer students are willing to study some of the biological sciences is the wretched literature they have to read!
>
> (British Ecological Society reader)

> ...from extensive experience as an editor, I can say that if the average standard of writing of papers submitted for publication were to rise to that of **any** of the examples given, I would think the millenium had arrived.
>
> (Biochemical Society reader)

Please make as much of this exercise as you can – it would save me countless hours if pseudo-intellectual gibberish were cast out.

(Institution of Chemical Engineers reader)

Texts used in the survey of readers' responses to various styles for a technical manual

What is the 'best' style for manuals?

Which of these six versions of a text do you consider the most acceptable as part of a **reference manual designed for programmers and field engineers?**

The text is about EDITMENU – a program that creates and maintains menu files, and adds new options to existing menu files.

It is desirable that there should be as much variety as possible in the order in which assessors read these texts. So please begin by reading the version with the letter closest to the initial letter of your surname.

For realism, these versions are based on genuine material; but I should like to stress that none of these versions is the responsibility of the originator(s) of the original text. I have changed the program names. I apologise in advance if my invented names inadvertently coincide with names already in use elsewhere.

Version M

EDITMENU operates on its own workfile. Accordingly, when you are maintaining a Menu file, none of your changes is made until either you finish your session or you access another Menu file. So the original files remains safe if, for some reason, your session does not end normally. EDITMENU does not recover any Menu file amendments that you made prior to a system or program failure.

If the file you are updating with EDITMENU is also the file being used by STARTSYSTEM, your updates will not affect STARTSYSTEM until STARTSYSTEM is terminated and re-started.

EDITMENU uses only the data file part of the indexed file pair. When your session ends, EDITMENU creates only the Menu data file. The corresponding keyfile is created by STARTSYSTEM (through SORTVALUES), as soon as it accesses the file. If you suspect incompatibility in the indexed files pair, you can remove the keyfile by running EDITMENU. This allows STARTSYSTEM to create a new keyfile.

Version T

EDITMENU operates on a workfile of its own. Thus, when maintaining a Menu file, the original remains unchanged until either the session finishes or another Menu file is accessed. This means the original file remains safe if the session does not terminate normally. EDITMENU does not perform recovery of Menu file amendments which were made prior to a system or program failure. If the file which is being updated via EDITMENU is also the file currently in use by STARTSYSTEM, any changes will not affect STARTSYSTEM until STARTSYSTEM is terminated and started up again.

EDITMENU uses only the data file part of the indexed file pair. On completion of a session, the program creates only the Menu data file. The corresponding keyfile is created by STARTSYSTEM (via SORTVALUES) when it first accesses the file. If the indexed file pair is suspected of being incompatible, running EDITMENU will cause the keyfile to be removed, allowing STARTSYSTEM to create a new one.

Version R

EDITMENU operates on a workfile of its own. So, when a Menu file is being maintained, the original of the file is not changed until the session finishes or until another Menu file is accessed. This preserves the original file if the session does not terminate normally. EDITMENU does not recover Menu file amendments that were made prior to a failure of the program or of the system.

If the file being updated by EDITMENU is also the file being used by STARTSYSTEM, the updates do not affect STARTSYSTEM until STARTSYSTEM is terminated and started up again.

EDITMENU uses only the data file part of the indexed file pair. When a session is complete, EDITMENU creates only the Menu data file. The corresponding keyfile is created by STARTSYSTEM (via SORTVALUES) when it first accesses the file. If the indexed file pair is suspected of being incompatible, the keyfile can be removed by the running of EDITMENU. This allows STARTSYSTEM to create a new keyfile.

Version Y

EDITMENU operates on a workfile of its own. So, when you are maintaining a Menu file, your original is not changed until either you finish your session or you access another Menu file. In this way, you keep your original file safe, even if your session does not terminate normally for any reason.

If your system breaks down or your program fails, EDITMENU does not recover any of the amendments you made before that happens.

If you are updating a file with EDITMENU, and that file is also currently being used by STARTSYSTEM, any changes you make will not affect STARTSYSTEM until you terminate STARTSYSTEM and then start it up again.

When you use EDITMENU, you work on only the data file part of the indexed file pair. When you complete your session, EDITMENU creates for you only the Menu data file. The corresponding keyfile is created for you by STARTSYSTEM (via SORTVALUES) when it first accesses the file. If you are suspicious about the compatibility of the indexed file pair, you can run EDITMENU and remove the keyfile. By doing that, you make it possible for STARTSYSTEM to create a new keyfile.

Version B

The operation of EDITMENU takes place on a dedicated workfile. Therefore, during the maintenance of a Menu file, no changes to the original file are effected until such time as the session comes to an end or the operator accesses another Menu file. The implication of this is that the original file will remain without alteration if, for some unforeseen reason, the operating session does not come to a normal termination. Amendments to a Menu file that were made before a system or program failure occurs are not recovered by EDITMENU.

If the file that is being subjected to update by EDITMENU is also the file currently being utilized by STARTSYSTEM, no effect on START-SYSTEM will be caused by any changes made by EDITMENU until after the termination and re-start of STARTSYSTEM.

Of the two parts of the indexed file pair, only the data file part is utilized by the Menu data file. Creation of the corresponding keyfile is accomplished by STARTSYSTEM (through SORTVALUES) when the file is first accessed by STARTSYSTEM. If a suspicion of incompatibility between the parts of the indexed file pair exists, removal of the keyfile can be achieved by the running of EDITMENU. This makes possible the creation of a new keyfile by STARTSYSTEM.

Version G

EDITMENU operation is on a dedicated workfile. Thus, during Menu file maintenance, no change in the original is performed until session termination or alternative Menu file access. Original file safety in the event of abnormal session termination is thereby ensured. Recovery of Menu file amendments made prior to a system or program failure is not an EDITMENU function.

If EDITMENU updating is being performed on a file that is also the file currently under utilization by STARTSYSTEM, no change is effected in STARTSYSTEM until termination and subsequent re-start of STARTSYSTEM.

EDITMENU utilizes only the data file portion of the indexed file pair. At session completion, EDITMENU file creation comprises the Menu data file only. Corresponding keyfile creation is by STARTSYSTEM (via SORTVALUES) on initial file access. Rectification of suspected incompatibility of the indexed file pair is possible by removal of the keyfile (by running of EDITMENU) thereby allowing new keyfile creation by STARTSYSTEM.

Results and comments from the survey of style for a technical manual

In making a fair assessment of responses to an extract from a reference manual, my main problem lay in the audience specified for it by its originators. The audience was 'programmers and field engineers'. Field engineers in the computer industry constitute a group that can be identified fairly easily. To identify 'programmers' is not so simple. By making enquiries in a few companies, I discovered that there are not many people whose job title is 'programmer'. In contrast, there are large numbers of people, many of whom do not consider themselves computer specialists, whose job entails the writing of programs, and who have to use reference manuals. I chose, therefore, to seek responses from two audiences whose work might reasonably involve them in reading, writing, editing, or using reference manuals for computers. One audience consisted of a 'mixed' group of participants on short courses on writing and editing; the other consisted of professionals in computing.

Responses from a 'mixed' audience

I run many short courses on writing and editing for people in industry, research centres, and government departments. The participants are mixtures of professional writers and editors, engineers, scientists, computer specialists and administrators. The courses are usually entitled 'Effective technical writing'. I am not able to obtain full details of education and occupation from the participants. I can say only that they are all involved in writing and reading technical documents in their professional lives.

During my courses, I have asked more than 1000 participants to judge the readability of six versions of an extract from a descriptive text. The content of the versions was held constant: only the style was changed. Readers were asked to judge the suitability of the text as part of a manual aimed at programmers and field engineers. (I hasten to say that I sought readers' opinions **before** we began to discuss style on the courses!) If you are a reader from the same broad group, and you have come straight from Chapter 1 to these Appendices, I should be interested to hear how your judgements compare with the judgements I have obtained so far.

My sample of 1084 respondents in this 'mixed' group voted for Version Y as best by a large majority. Version Y is one of two versions that use a second-person (*you*, *your*) style; Version M is the other second-person version. Together, those two versions took 66.5% of the votes:

Table 1 Results from the 'mixed' group

Version	Best version (%)	Worst version (%)
B	3.0	45.0
G	3.8	36.0
M	22.9	2.8
R	17.4	1.3
T	9.3	9.2
Y	43.6	5.6
	100.0 (n = 1084)	99.9 (n = 996)

Versions B and G bore the brunt of voters' disapproval. The longest of the six versions, Version B, was declared worst of all. Significantly – since we are often told that brevity is the essence of good technical writing – the shortest version, Version G, was almost as unpopular.

I am well aware that my opinion-seeking activities were not rigorous scientific research. I **ought** to have manipulated just one variable in each version, and held all the other variables constant. That would have been an impossible task; and even if it had been possible, it would have required respondents to read dozens of versions of the text, which would have been impracticable. Accordingly, I claim only that the results from the 'mixed' group of participants in my courses give interesting information about what those participants considered good style for a short descriptive text.

Responses from professionals in computing

Nevertheless, although I have always restricted my claims carefully, many readers have been reluctant to accept the implications of the results. I have been challenged on many occasions with the argument: 'It's all very well for **us** to vote this way, but programmers and field engineers in general would not like Versions Y and M; they would think them too casual; **they** prefer impersonal writing'.

I recognize the power of mythology, and the influence of the spectral group 'they'! Accordingly, I obtained a sample of 358 opinions from professionals in computing – people who were either working in computing organizations, or who were readers of a computing journal. From this group, the results were:

Table 2 Results from professionals in computing

Version	Best version (%)	Worst version (%)
B	5.0	37.6
G	3.6	47.2
M	17.9	3.4
R	20.1	1.1
T	13.7	3.4
Y	39.7	7.3
	100.00 (n = 358)	100.00 (n = 358)

In this sample, Y was still the most popular version; but the pattern of judgements did differ slightly from the pattern produced by my short-course participants. Version R rose above Version M in rank. All the impersonal texts, especially Version R and Version T **did** gain more votes. That lends support to those who fear that programmers and field engineers prefer a heavier, impersonal style. But Y and M still took more than 57% of the votes. If we ignore votes for B and G (which are not really serious contenders for the title 'best version'), Y and M together beat R and T together by 58% to 34%.

If you have never before seen results from readability tests, you may be thinking that the most remarkable outcome of the over-all survey is that some votes were cast for **all** the versions as 'best', and some votes were cast for all the versions as 'worst'! Responses to style are curious mixtures of reason and emotion. Witness these comments from respondents:

'Use of *you* is important, especially when discussing compatibility of file index – unless it is clear that 'you' can be suspicious, you might expect a system message 'suspected indexed file pair incompatibility' or similar.'

'I think the use of the personal pronoun is unnecessarily patronizing – at any level. I do not like 'so' at any price.'

'2nd person is much more encouraging to read.'

'(2nd person) is. . .for the wally end-user.'

'. . .user-chatty is a desirable quality in both on-line and paper text.'

'. . .the second-person grammer (sic) is tiresome.'

'T and Y both work well. As a technician, I prefer T for its impersonal style. Y is probably better for an end-user than a support person.'

'Y is the best written..., but the personal form of address could be in the way for an experienced person.'

(from an experienced programmer!)

It would be helpful to teachers of writing if we could produce absolute evidence that one style is more readable and more acceptable than another. Unfortunately, I do not know of any survey of responses to styles that has produced unanimous voting in favour of one 'version' out of many. The most we can do is obtain a large number of judgements, and make known what is thought 'best' by the majority of readers in any specialist environment. I hope you will think these results are a useful step in that direction.

The differences among the six versions are not simply matters of overall length, or of the use of impersonal or personal structures. It is convenient, though an over-simplification, to see the 'linguistic variables' that are manipulated as a set of contrasts:

Wholly impersonal	vs	partly impersonal, partly second-person
Structures heavily 'noun-centred' (heavy 'nominalization')	vs	structures mainly 'verb-centred'
Verbs mainly passive	vs	verbs mainly active
Vocabulary unnecessarily 'grandiose'	vs	vocabulary as 'plain' as possible
Liberal paragraphing	vs	less paragraphing
Sentences often long and complex	vs	sentences mainly short and manageable.

Second-person style

The most obvious difference produced by manipulation of 'linguistic variables' in the six versions is that created by introducing second-person style ('you' and 'your') in two, in contrast with the impersonal style of all the others:

Version B: Therefore, during the maintenance of a Menu file, no changes to the original file are effected until such time as the session comes to an end or the operator accesses another Menu file.

Version G: Thus, during Menu file maintenance, no change in the original is performed until session termination or alternative Menu file access.

Version M: Accordingly, when you are maintaining a Menu file, none of your changes is made until either you finish your session or you access another Menu file.

Version R: So, when a Menu file is being maintained, the original of the file is not changed until the session finishes or until another Menu file is accessed.

Version T: Thus, when maintaining a Menu file, the original remains unchanged until either the session finishes or another Menu file is accessed.

Version Y: So, when you are maintaining a Menu file, your original is not changed until either you finish your session or you access another Menu file.

These short extracts demonstrate most of the advantages that accrue from use of a second-person style. Perhaps the most important is greater explicitness. Versions M and Y make clear **who** accesses another menu file: they make plain that **you** do so, not the machine or another user.

Elsewhere in the texts, the same advantage appears. As one of the respondents commented:

'Several of these examples do not make clear whether the program automatically detects "suspected incompatibility" or whether the user must work it out.'

(Programmer or field engineer)

The use of *you* removes the uncertainty.

Introduction of *the operator* or *the user*

It is possible to avoid *you* by supplying another term, such as *the operator* or *the user*; but this sets up a pretence that three parties are involved in the discussion: the writer, the reader, and a third party – the operator.

Of course, there may be occasions when the reader of the manual is not the person who normally operates the machine; but it seems reasonable to write most of the time as though you were addressing the operator or potential operator. Most readers mentally put themselves in the place of the operator while they read the text, so it is helpfully direct to address them directly as *you*. As one respondent commented:

'M is consistent with my mental picture'

'Sexist' writing

A major disadvantage of using *the operator* or *the user* is that it often leads to 'sexist' writing. Once you have begun with *The operator...*, you are obliged to use *he, she* or *he/she* when you need a pronoun. For example, note what would have happened if I had begun the previous sentence with *Once a writer...*:

> Once a writer has begun with *The operator...*, he (she, he/she) is obliged to use *he, she,* or *he/she...*

Certainly, it is usually possible to recast your statement so that you can avoid accusations of sexism, or the clumsiness of repeated use of *he/she*. In the six versions, M and Y succeed by using *you*:

> ...when you are maintaining a Menu file...

R succeeds by using a passive-verb construction:

> ...when a Menu file is being maintained...

T avoids sexism, but lapses into ungrammatical expression (explained in the next section). B and G also succeed in avoiding sexism, but at the cost of using 'noun-centred' style (discussed in the next section but one).

Unfortunately, though they succeed in avoiding sexism, the avoiding tactics used by B, G, R, and T introduce ambiguity: they all leave uncertainty about who does the maintaining.

Ungrammatical constructions: misrelations

A particular drawback to impersonal thinking is that it tempts writers to write ungrammatically – to misrelate preliminary participial constructions. In English, we have a convention that a preliminary clause centred on an *-ing* participle relates to the first noun or pronoun in the following main clause. We should therefore write:

> After loading the cassette, you are asked to choose one of three options from the main menu...

not:

> After loading the cassette, the main menu asks the user to choose one of three options...

Unfortunately, in its effort to remain entirely impersonal, Version T slips into one of these misrelated participial constructions – a grammatical error that creates a nonsensical statement:

...when maintaining a MENU file, the original remains unchanged...

For further disscussion of this point, see Chapter 2, section 2.13, pages 73 to 77.

'Noun-centred' style vs 'verb-centred' style

Another disadvantage to impersonal style is that it encourages writers to write in a 'noun-centred' way instead of in a 'verb-centred' way: that is, to write:

...during Menu file maintenance, no change in the original is performed...

instead of:

...when you are maintaining a Menu file, your original is not changed...

Excessive use of 'abstract noun + passive verb' constructions (excessive 'nominalization') is a major cause of the long-winded heaviness of much technical writing. If you begin with the expression *The operation of EDITMENU*..., you are obliged to complete your statement with a colourless 'general purpose' verb such as *takes place, is carried out, is effected, is achieved,* or *is accomplished*. At least five words have to be used, where the two words *EDITMENU operates*... would have been entirely adequate.

The two-word version also creates a statement that has a more explicit action expressed by its heart – the verb: in the two-word version, some *operating* is taking place; in the five-word version, some *taking place* is taking place!

I acknowledge that there are many occasions in technical writing when it is desirable to focus on an abstract notion, and to sustain that focus from sentence to sentence. In such circumstances, it is valuable – even essential – to use passive constructions. For example:

...requires an increase in the flexibility of the wing. The necessary flexibility is achieved by constructing...

Frequently, however, writers seem to get fixed into a mind-set that turns most sentences – unnecessarily and disadvantageously – into roundabout 'abstract noun + passive verb' constructions:

...no effect on STARTSYSTEM will be caused by any changes made by EDITMENU...

(changes made by EDITMENU will not affect STARTSYSTEM)

If EDITMENU updating is being performed...
(If EDITMENU is being updated)

In general, 'verb-centred' structures give writing greater vigour and readability than is produced by 'noun-centred' structures. The two examples immediately above show that this is true whether the verb form is active (*will not affect*) or passive (*is being updated*).

However, as every textbook on technical writing declares, it is helpful to use active-voice constructions rather than passive-voice constructions wherever possible. A preponderance of passive constructions usually lengthens a text:

Write You can remove the keyfile by running EDITMENU...

Not Removal of the keyfile can be achieved by the running of EDITMENU...

For further discussion of this point, see Chapter 2, sections 2.10 and 2.11, pages 50 to 51 and 52–9.

Mixtures of 'good' and 'bad' features in the six versions

It would have been unfair to have put all the 'good' features of style in just one or two versions, and all the 'bad' features in the others. For this reason, not only do M and Y have mainly 'good', active verbforms: R and T have them, too.

Similarly, not only does the impersonal version R contain a 'bad' sentence, starting *So,...*: the second-person version M has one, too.

Over the years, I have found that many writers and readers react vehemently against the use of 'so':

'I do not like "so" at any price'
(Professional technical writer/editor)

'Any version which has a sentence beginning with the word "so" is excluded. I am sure it is bad English.'
(Programmer/field engineer)

It seems that teachers of English in schools have left many people with guilt feelings about beginning a sentence with *so*, though other adverbs and conjunctions are apparently acceptable. In technical documents, it seems permissible to begin sentences with *accordingly, thus, hence, therefore, equally, again, similarly, consequently, likewise,...*; but *so* produces a knee-jerk adverse response.

To suggest that this response is irrational is not to suggest that it should be ignored. If our documents are to be effective, they must produce a favourable response from as many readers as possible. If we know that certain features of style will be irritants to some of our readers, it is wise to find another way of expressing our meaning – without, of course, distorting what we want to say.

I included *So,...* constructions in version M and R to see if those 'bad' features of style would weigh so heavily as to make readers reject M and R entirely. The voting papers make clear that some voters were put off by the presence of the *So,...* constructions; but the total scores for M and R show that many readers either were not concerned about the sentences starting with *So,...*, or felt that the good qualities of those two versions far outweighed their bad qualities.

Jargon and 'plain' vocabulary

Versions R and T were the most popular impersonal versions. Like M and Y, they do not use unnecessarily 'grandiose' language. They say that EDITMENU operates on *a workfile of its own*, not on *a dedicated workfile* (B and G). R and T have *being used* and *in use* instead of *being utilized* (B) and *under utilization* (G). R has EDITMENU *does not recover ...amendments* instead of G's *Recovery...is not an EDITMENU function*.

Note, however, that all **necessary** jargon is present in all six versions. The special terminology of computing cannot be avoided if we are to write accurately and adequately: but we **can** keep our writing more direct and comfortable if we avoid the temptation to use fashionable, inflated vocabulary like *enhance* (does that mean 'increase', 'improve', or 'add'?), *power up* (does that mean the same as 'switch on'?), or *terminate* (does that mean 'finish normally, end' or 'cut off, stop'?).

A particular danger of the habit of using 'grander' language than is necessary is that we are tempted to use true technical terms casually and inaccurately. The writer who wrote:

...de-energize the primary power source...

did not mean *de-energize*: he meant 'switch off'. His use of the grandiose expression the *primary power source* implied that there was a secondary source. There was not. He meant 'Switch off the electricity'.

Length and complexity of sentences

In general, M, R, T and Y seem reasonably direct, active, 'plain' statements; B and G are less accessible. Is the greater readability of M, R, T and Y derived from differences in the length and complexity of sentences?

The sentence- and paragraph-structure of the versions is:

Table 3 Sentence- and paragraph-structures in all versions

	B	G	M	R	T	Y
No. of paragraphs	3	3	3	3	2	4
No. of sentences	9	9	10	10	9	10
No. of words	209	153	158	164	160	194
Average length of sentences	23.2	17.0	15.8	16.4	17.8	19.4

It is interesting that the most popular version overall, Y, is not much shorter than the least popular version over-all, B. Also, the shortest version, G, was not well received.

There is no doubt that the length and the complexity of sentence structures are major influences on the ease with which readers are able to assimilate information; but neither length nor complexity alone necessarily causes difficulty. Certainly, version B suffers from having a sentence that contains 41 words; but Y has one that contains 36 words, yet Y does not suffer so badly. It is the **combination** of excessive sentence-length and the redundancy of much of B's wording that makes most of us feel that B is too loquacious. Similarly, the average length of sentences in T is greater than the average in G; but it is the greater use of complex noun groups within G's sentences that makes most of us feel that G is denser and more difficult to penetrate than T.

Cumulative effects of writers' choices

By now, those of you who did not vote for M or Y will be feeling that my analysis is heavily biased. I acknowledge that I prefer the style of M and Y, but I would emphasize that I have simply tried to identify what made M and Y come out top in my poll of readers.

In my view, the most important finding to emerge from any analysis of response to styles is that no single linguistic feature is responsible for a text seeming good or bad: readers' responses are affected mainly by the cumulative effect of the linguistic choices made by writers. By itself, the use of an unnecessarily grandiose term *dedicated* seems of little significance; by itself, the use of a roundabout, noun-centred construction *no change in the original is performed* instead of the more direct *the original is not changed* seems of little significance; by itself, the unnecessary use of the long-winded *until such time as* instead of *until* seems of little significance. Cumulatively, however, a writers' indulgence in these undesirable features builds a sense of struggle and resentment in his or her readers.

Further debate

More debate about the versions is possible. Surely, *Menu* should not have a capital *M* in any version? How important is punctuation in showing up the meaning? Are parenthetical insertions necessarily unmanageable? Are bits of Latin (*via*) really necessary in modern text? Do the width of text and/or the frequency of word-breaking influence your judgement as you read a text?

I hope this analysis will be useful in promoting debate. I shall be pleased if you are able to obtain more judgements on the texts (**before** the readers have read the six versions!). Let me know if I can help in any way.

Somehow, we have to improve the overall standard of writing in manuals and on-screen texts. In particular, we have to discourage the style of Version G, which produced these comments from programmers:

'I concentrated on the "pompous" style rather than the content.'
'I have 30 years experience of (mis)reading such stuff.'

But it is salutary to know that professionals in science and technology do not feel satisfied with the general standard of the writing they have to read. Many of the programmers and field engineers made a general comment that can be summed up in one quotation:

'I thought they were all pretty poor – fully down to current tech. manual standards.'

References

1. Tinker, M.A. (1963) *Legibility of Print*, Iowa State University Press.
2. Turnbull, A.T. and Baird, R.N. (1968) *The Graphics of Communication: typography, lay-out and design*, Holt, Rinehart and Winston.
3. Macdonald-Ross, M. and Smith, E. (1977) *Graphics in Text: a bibliography*, IET Monograph No. 6, Open University.
4. Pickering , G. 'Language, the lost tool of learning in medicine and science', *The Lancet*, 15 July 1961, p. 116.
5. Fowler, H.W. (1962) *Modern English Usage*, (second edition revised by Sir Ernest Gowers), Oxford University Press, p. 461.
6. Quirk, R., Greenbaum, S., Leech, G. and Svartvik, J. (1989) *A Comprehensive Grammar of the English Language*, Longman, p. 1243.
7. Gunning, R. (1962) *More Effective Writing in Business and Industry*, Farnsworth Publishing Inc., pp. 5–29.
8. Gowers, E. (1962) *The Complete Plain Words*, Pelican Books, p. 136.
9. Quirk, R., Greenbaum, S., Leech, G. and Svartvik, J. (1972) *A Grammar of Contemporary English*, Longman, p. 807.
10. Good, I.J. (1962) *The Scientist Speculates*, Heinemann, pp. 52–3.
11. Kirkman, J. (1989) *Full Marks: Advice on Punctuation for Scientific and Technical Writing*, JKCC, PO Box 106, Marlborough, Wiltshire SN8 2RU.
12. Fowler, H.W. (1962) *Modern English Usage*, (second edition revised by Sir Ernest Gowers), Oxford University Press, pp. 587–92.
13. Gowers, E. (1962) *The Complete Plain Words*, Pelican Books, pp. 236–60.
14. Carey, G.V. (1976) *Mind the Stop*, Penguin Books.
15. Partridge, E. (1964) *You Have a Point There*, Hamish Hamilton.
16. Carey, G.V. (1976) *Mind the Stop*, Penguin Books, p. 22.
17. Shneidermann, B. (1982) *Directions in Human-Computer Interaction*, Ablex.
18. Shneidermann, B. (1987) *Designing the User Interface*, Addison-Wesley.
19. Horton, W. (1990) *Designing and Writing Online Documentation*, Wiley.
20. Barrett, E. (ed. 1988) *Text, ConText and HyperText*, MIT Press.
21. Royal Air Force, Central Servicing Development Establishment.

Crown Copyright, reproduced by permission of the Controller of Her Majesty's Stationery Office.

22. Matthies, L. (1977) *The New Playscript Procedure*, Office Publications Inc.

23. Kirkman, J., Snow, C. and Watson, I. (1978) *Controlled English in International Documentation*, Proceedings of the 39th Congress of the International Federation for Documentation, Edinburgh, 25–28 September.

24. *Ibid.*

25. Kirkman, J., Snow, C. and Watson, I. (1978) *Controlled English as an Alternative to Multiple Translations*, IEEE Transactions on Professional Communication, Vol. PC-21, No. 4, pp. 159–61.

26. Department of Trade (1974) *Standard Marine Navigational Vocabulary*, Merchant Shipping Notice No. M702, London.

27. Ehrlich, E. and Murphy, D. (1964) *The Art of Technical Writing*, Bantam Books, p. 113.

28. Little, P. (1965) *Communication in Business*, Longmans, p. 41.

29. Gloag, J. (1950) *How to Write Technical Books*, Allen and Unwin, p. 30.

30. Graves, H.F. and Hoffman, L.S.S. (1965) *Report Writing*, Prentice-Hall, p. 26.

31. Emberger, M.R. and Hall, M.R. (1955) *Scientific Writing*, Harcourt Brace, p. 157.

32. Salomon, L. (1966) *Semantics and Common Sense*, Holt, Rinehart and Winston, p. 14.

33. Empson, W. (1961) *Seven Types of Ambiguity*, Penguin Books, p. 1.

34. Flood, W.E. (1960) *Scientific Words, Their Structure and Meaning*, Oldbourne, p. viii.

35. Waismann, F., 'Verifiability', in G.H.R. Parkinson (ed. 1968) *The Theory of Meaning*, Oxford University Press, p. 38.

36. Quine, W.V.O. (1960) *Word and Object*, MIT Press, pp. 125–6.

37. Kirkman, J. 'Command of vocabulary among university entrants', *Educational Research*, Newnes Ltd., Vol. 9, No. 4, 1967, p. 151.

38. Flood, W.E. (1957) *The Problem of Vocabulary in the Popularisation of Science*, Oliver and Boyd, 1957, p. 87.

Index